AF531447

ELEMENTARY CHEMISTRY

By

Dr. R.K. Sharma

DISCOVERY PUBLISHING HOUSE
NEW DELHI-110002

First Published-2007

ISBN 978-81-8356-212-6

Published by

DISCOVERY PUBLISHING HOUSE
4831/24, Ansari Road, Prahlad Street,
Darya Ganj, New Delhi-110002 (India)
Phone: 23279245 • Fax: 91-11-23253475
E-mail: dphbooks@rediffmail.com
dphtemp@indiatimes.com

Printed at:

Sachin Printers, Delhi

Preface

This book has been designed to cover the syllabus of Inorganic Chemistry required for the B.Sc./B.Sc. Hons./M.Sc. students of the Indian Universities. I have compelled all the questions asked so far in different universities as well as C.C.S. University Meerut. I have arranged the subject matter in a continuous manner. Special emphasis has been laid on fundamental concept of the topics.

Care has been taken to make the treatment of the subject simple and accessible to the average students. The gradation of the solved and unsolved questions has been facilitated from simple to difficult. It is believed that the book in the present form will be found to be useful by the student community and the teaching fraternity alike.

Suggestion for the improvement of the book will always be most welcome.

Author

CONTENTS

Composition of a Compound, Chemical Equations and their Balancing, Balancing of Chemical Equations, Partial Equation Method, Limiting Reagent, Calculations using Chemical equations, Problems Based on Mass-Volume, and Volume-Volume Relationship, Concentration of Solutions and Problem based on them, Normality and Molarity Equations, Conceptual Corner for Competitive Examinations, Important Formulae, Solved Examples.

1

BASIC CONCEPT OF CHEMISTRY

CHEMISTRY AS A NATURAL SCIENCES

Chemistry is concerned with the knowledge of the natural world. Hence, it is usually called a *Natural sciences*. The separation of the natural sciences into physical and biological sciences and physical sciences into physics and chemistry divides a larger body of knowledge into more manageable branches. However, the concepts, techniques and applications of the various sciences are interdependent. Most students regardless of the area of natural science they wish to pursue need an introduction to the principles and applications of chemistry as a foundation for their speciality.

CHEMISTRY AS A TECHNOLOGY

Chemistry is a very interesting subject. Many products of the "test tube" have become common household items—plastics, synthetic fibres, synthetic rubber, deodorants, soaps, body and hair sprays etc., are all the products of chemistry. The chemists in our modem society are busy in producing novel materials—wonder drugs, smart materials etc. Today about 80% of the newly trained chemists enter careers in chemical industry.

CHEMISTRY IN OUR DAILY LIFE

The things like glass window, nylon, clothes synthetic rubber, detergents, soap has its origin in chemistry. We observe that almost everything around is either a product of some chemical industry or has its origin in chemistry.

Much of our food, medicines and drugs are also prepared in chemical laboratories. Every day, more and more new drugs are discovered for the benefit of mankind. In fact, people everywhere are getting more and more curious to know which product is the best to suit their needs.

Did you ever stop to consider what this world would be if all the things we see around us were not there?

How has Chemistry transformed this planet of ours so profoundly and in such a relatively brief time? To understand this tremendous change, one should have some basic knowledge of chemistry and its applications.

CHEMISTRY AS AN INTELLECTUAL DISCIPLINE

In chemistry, there is a huge body of knowledge, facts, theories and applications already worked out. This gives an impression that all answers in chemistry have been found and there is nothing new to discover. This impression is not true because even today more than 3 lakh new compounds are made annually and about 2 million enterics are reported in chemical literature every year.

Thus, more new chemistry than ever is now being discovered these days. As more and more facts are discovered, more and more theories are developed, and more and more applications are found, there are indications that chemistry continues to grow as an intellectual discipline. Many more areas of chemical research leading to many interesting developments are sure to come up in near future.

CHEMISTRY IN WORLD PROBLEMS

Some of the priority problems facing the human race in which chemists are expected to make significant contributions towards their solutions are listed below.

The food-population problem : In parts of the world where population is increasing much more rapidly than the food production famine-like conditions exist. The Green Revolution has partly relieved us of this problem but more vigorous efforts are needed to find long lasting solution to this problem. Chemists are involved in the development of the following:

(a) More suitable disease and pest controls and" fertilizers.

(b) Acceptable controls of human fertility.

(c) Production of synthetic foods based on non-agricultural raw materials.

Chemists are developing synthetic plant hormones and growth regulators to increase plant yield. Chemistry also plays an important role in the preservation of food.

(d) *Environmental protection :* Finding an acceptable solution to the air and water pollution is a continuing challenge. Disposal of solid and liquid wastes is the biggest problem for urban authorities.

The use of fossil fuels has contributed to the greenhouse effect and polluted the air. The exhaust gases from automobiles contain carbon monoxide, unburnt hydrocarbons, oxides of sulphur and nitrogen and lead particulates in addition to carbon dioxide.

Carbon monoxide is more dangerous pollutant of the air. Los Angeles smog is caused by the excessive emission of exhaust gases from automobiles.

To reduce air pollution, chemists must find new fuels, and new catalysts. Greater reuse and recycling of materials is a way to find a solution for garbage disposal, at least to some extent. Recycling of paper products will save trees from large scale cutting. Glass bottles and aluminium cans can also be recycled easily.

The combustible materials in the garbage can be used to generate electricity. Many countries have developed garbage disposal procedures on the basis of recycling. A lot of work is being done in this direction at many places in the world.

Oceans as the sources of raw materials : We already obtain some food and a few minerals—magnesium, bromine, iodine etc., from the oceans. Oceans are found to be storehouse of many kinds of raw materials. It is due to the efforts of scientists that we have started exploiting the vast resources of oceans. Chemists are involved in successful recovery of usable water and many minerals from the oceans.

Biochemical processes : We have learnt a lot about metabolism of food, photosynthesis, and other biological phenomena by applying the principles of chemistry to biological processes. Recently, a new area of biochemical research has opened up the possibilities of even playing with the genetic code of human beings.

In brief, our life cycle in principle is a sort of chemical clock, ticking with the highest degree of precision. Chemists has a very important role to play in understanding these processes.

New sources of energy : Coal, petroleum, and natural gas may last another fifty years. Production of nuclear energy is also limited to the availability of the fissionable material. In the last few decades, attention is focussed on the renewable resources of energy — wind energy, solar energy, tidal energy etc. Each of these newer sources of energy will require materials and procedures for producing and transferring energy from its source to the user. Chemists can help a lot in developing suitable materials needed in such sources of energy. Recently, a German research group has developed a catalyst to split water into hydrogen and oxygen with the help of sunlight. Fuel cell technology is just ready to come out for common man's use.

New materials : With the increased demand for new products, requirement for new materials has also increased. With advances in technology, new materials such as super-conductors, high field magnetic materials, high temperature plastics and structural polymers have been developed and put to use. Many more materials for some specialized uses are being developed and studied. For example, conducting polymers, smart materials, materials based on nano-structures are at various stages of development. All these materials can be prepared only with a considerable knowledge and experience in chemistry. Chemists have already produced synthetic blood and skin. From all what has been described above, it is clear that chemistry as a science is very much with us today and its future holds a bright promise of revealing much more to come.

MEASUREMENT IN CHEMISTRY

Chemistry is an experimental science. Like all branches of science, experiments in chemistry involve both observation and collection of quantitative data. The data collected for a wide variety of processes are processed and analysed to establish certain relationships. Such relationships, trends and similarities help the scientists to postulate theories. Thus, it becomes quite obvious that the measurement is a very important aspect of any experimental science. In this unit, we describe the process of measurement and the correct reporting of the results obtained through such measurements.

MEASUREMENT OF A PHYSICAL QUANTITY

The measurement of any physical quantity involves *the comparison of its magnitude (size) with a standard of the same kind.* The standards are fixed arbitrarily, but there is always a logic behind such a choice. The arbitrarily fixed standard is called a unit. For example, for measuring length, a standard unit of metre is used, for measuring mass, a unit called kilogram is used.

Generally, these standard units are prescribed by the International *Bureau on Weight and Measures*. For example, a kilogram (unit of mass) is defined as follows:

The mass equal to the mass of a cylinder of platinum-irridium alloy kept by the International Bureau on Weights and Measures at Paris is called a kilogram.

Thus, to express any measurement one should know the following:

(i) The *unit* in which the quantity is measured

(ii) The *numerical value* of the measure. This numerical value gives the number of times the unit is contained in it.

For example, while measuring the length of a piece of cloth, we should know,

(i) The unit in which the length is measured, *i.e.*, whether the length is measured in metres or feet?

(ii) The number of units in the whole length of the piece of cloth, *i.e.*, the number of metres or feet in the whole length of the piece of cloth. For example, the length of the given piece of cloth is 2.5 m means that 2.5 units (metre) are contained in the length of the cloth piece: the unit in this case being equal to 1 metre.

The numerical value of any measurement depends upon the following factors.

(i) The *reliability* of the instrument used in the measurement

(ii) The *judgement* of the person making the measurement.

A reliable measurement must be reproducible. Reproducibility means, that if a measurement is made repeatedly, then the various measurements should either be the same or fall within a certain limit of uncertainty.

Precision and Accuracy of a Measurement

The limit of the reproducibility of any measurement is known as its precision. A measurement may be absolutely precise, or it may have some uncertainty depending upon the least-count of the instrument and the observation power of the person making the measurement. For example an exact number, *e.g.*, number of eggs, number of men etc., is absolutely precise, but the measured mass of an object is not absolutely precise. The same object may weigh differently on different scales. The precision in the measurement of mass of an object depends upon the,

(i) Sensitivity of the weighing scale (the balance),

(ii) The observation power (the skill) of the person making the measurement.

The precision of any measured quantity is nothing to do with the accuracy of results. The accuracy of any measurement may be defined as follows:

The degree of conformity of a measure to its true value is termed its accuracy.

To make you understand, we describe the following example.

You go to the Shopping Mall, and get yourself weighed on different weighing machines. You will observe that your reported weights differ from machine to machine, viz., as follows.

Machine	:	A	B	C	D	E
Mass (kg)	:	38	39	42	39	41

Thus, each machine gives different mass value.

Then, you weigh yourself on the machine E, four times. Every time the machine E gives the same mass (say 41 kg).

Now you weigh yourself on a standardised doctor's weighing machine. Suppose, the doctor's weighing machine records your mass as 40 kg. Let us take this as your true (actual) mass.

All the four weighing on machine E are precise, but not accurate: the accurate value is 40 kg. Machines A to E give masses which neither agree with each other, nor equal to the accurate mass. Each of these masses is uncertain by certain amount to the actual (true) value.

The uncertainty in any measurement is shown as a limit of fluctuation as $\pm\ \delta x$ alongwith the number.

For example, 160 ± 1 cm means that the true value may lie between 159 (= 160 ± 1) cm and 161 (= $160 + 1$) cm.

INTERNATIONAL SYSTEM OF UNITS—THE SI UNITS

We generally deal with a large number of physical quantities. Each physical quantity requires a separate unit for its measurement. As a result, we have to deal with a large number of units of different kinds. It is not practical to deal with as many units as there are physical quantities.

With the advances in science, the need of a uniform system of measurements was felt. To make measurements more scientific, convenient and uniform, the French Academy of Science in 1791 devised a *metric system of measurement based on the decimal system.*

In October 1960, the XIth General Conference of Weights and Measures (*Conference Generate des Poids at Measures*), adopted an International System of Units. The units in this system are known as SI units. The abbreviation SI comes from the French word, *Systeme Internationale d' Unites*). The SI units are now accepted universally for use in the scientific literature.

The Base SI Units

In international system of units, there are seven base units (Table 1.1).

Table 1.1 : The seven base SI units

Quantity	*Symbol for quantity*	*Name of SI unit*	*Symbol for SI unit*
1. Length	I	metre	m
2. Mass	m	kilogram	kg
3. Time	t	second	s
4. Electric current	I	ampere	A
5. Thermodynamic temperature T	kelvin	K	
6. Amount of substance	n	mole	mol
7. Luminous intensity	I_v	candela	cd

The definitions of these seven base units are given below:

1. *Length* : The base unit of length or distance is metre (m). A metre is defined as the distance travelled by light in vacuum during 1/299 792 458 second.

2. *Mass* : The unit of mass is kilogram. A kilogram (kg) is equal to the mass of platinum-indium cylinder (known as the standard kilogram) kept at the International Bureau of Weights and Measures at Sevres, France.
3. *Time* : The base unit of time is second. *A second* (s) is defined as the duration of 9,192,631,770 periods of the radiation corresponding to the transition between two hyperfine levels of the ground state of the cesium-133 atom.
4. *Temperature* : The base unit of temperature is kelvin (K). The thermodynamic scale on which temperature is measured has its zero at absolute zero, and has a fixed point corresponding to 273.15 K at the triple point of water.
5. *Electric current* : The base unit of electric current is ampere. Ampere (A) is defined as the magnitude of the current that when flowing through each of the two long parallel wires of length equal to 1 m, separated by 1 metre in free space, results in a force of 2×10^{-7} N between the two wires.
6. *Amount of substance* : The base unit of amount is the mole. The *mole* (n) is defined as the amount of any substance which contains as many elementary units as there are atoms in exactly 0.012 kg (*i.e.*, 12 g) of $^{12}_{6}C$ (carbon atom of mass number equal to 12).
7. *Luminous intensity* : The base unit of luminously is *candela.* Candela is defined as the luminous intensity of 1/600000 of a square metre of a radiating cavity at the freezing temperature of platinum (2042 K).

Conventions in Writing SI Units

The following conventions (rules) should be strictly followed while writing SI units.

(i) While writing a unit, only its singular form is used. For example, the mass of any substance should be written as kg not as kgs. Distances should be expressed in km not in kms, *e.g.*, ten kilogram is written as

10 kg	10 kgs
(correct)	(incorrect)

(ii) The abbreviation of any unit does not have full stop (.) at the end, unless it appears at the end of a sentence. Therefore, no full stop is placed either in between or at the end of the symbol of the unit. For example,

kg	kg	cm	cm.
(correct)	(incorrect)	(correct)	(incorrect)

(iii) One space is left between the last digit of a numeral and the symbol of the unit. For example,

10 m	10m	5 g	5g
(correct)	(incorrect)	(correct)	(incorrect)

(iv) Words and Symbols should not be mixed. For example, kilogram per cubic metre kg/m^3 kilogram/m^3 kg per cubic metre

kilogram per cubic metre	kg/m^3	kilogram/m^3	kg per cubic metre
(correct)	(correct)	(incorrect)	(incorrect)

(v) With numerals, the symbol of the unit should be written. For example,

100 cm	one hundred cm	100 centimetres
(correct)	(incorrect)	(incorrect)

(vi) The names of the units derived from people's names are written in small letters. For example, the unit named after Joule is written as joule, (symbol: J). The unit of force named after newton is written as newton, (symbol: N).

(vii) The sign of degree (°) is not written when Kelvin scale is employed. For example, temperature of 273 kelvin is written as 273 K and not 273°K.

273 K	273°K
(correct)	(incorrect)

Prefixes in SI Units

The metric system has been adopted almost around the world. But, over the time, a number of other units (multiples and submultiples of the SI units) have come into use. This has happened because the SI units of some physical quantities are either too big or too small. The units smaller or bigger than those based on metric system are related to the metric unit by the powers of 10. Different powers of 10 are indicated by prefixes.

For example, one hundredth pan (10^{-2}) of a metre is called centimetre: centi means one-hundredth. A list of standard prefixes is given in Table 1.2.

Table 1.2 : The standard prefixes for obtaining submultiple and/ or multiples of metric units.

Multiple	*Prefix*	*Symbol*	*Submultiple*	*Prefix*	*Symbol*
1018	exa	E	10^{-1}	deci	d
1015	peta	P	10^{-2}	centi	c
1012	tera	T	10^{-3}	milli	m
109	giga	G	10^{-6}	micro	^
106	mega	M	10^{-9}	nano	n4
103	kilo	k	10^{-12}	pico	P
102	hacto	h	10^{-15}	femto	f
10	deka	da	10^{-18}	atto	a

DIMENSIONS AND DIMENSIONAL ANALYSIS

The physical nature of a quantity is denoted by its dimensions. The relationship between the unit of any physical quantity to the fundamental units (length, mass and time etc.) is indicated by the dimensions of the unit concerned.

- A numerical quantity has no dimensions.
- The dimension of length, mass and time are denoted by the letters L, M and T respectively.
- The dimensions of a physical quantity are denoted by the notation [......]. For example, the dimensions of velocity v, are denoted by [v], the dimensions of area, A, by [A] etc.
- The dimensional equation describes the relationship between the derived unit and the base units of length, mass and time. The dimensional equations help us in the conversion of units. The terms on both sides of the equation must have the same dimensions.
- Dimensions can be treated as algebraic quantities, that is quantities can be added or subtracted only if they have the same dimensions.

Dimensions of some simple physical quantities are given in Table 1.3.

Table 1.3 : Dimensions of some simple physical quantities.

Quantity	*Definition*	*Dimensions*
Area	Length × Breadth	$L \times L = L^2$
Volume	Length × Breadth × Height	$L \times L \times L = L^3$
Velocity	$\frac{\text{Dis tan ce}}{\text{Time}} = \frac{\text{Length}}{\text{Time}}$	$\frac{L}{T} = LT^{-1}$
Laws of Chemical Combination	$\frac{\text{Change in velocity}}{\text{Time}} = \frac{\text{Dis tan ce/Time}}{\text{Time}}$	$\frac{L/T}{T} = LT^{-2}$
Force	Mass × Laws of Chemical Combination	$M \times LT^{-2} = MLT^{-2}$
Energy	Force × Distance	$MLT^{-2} \times L = ML^2T^{-2}$
Pressure	$\frac{\text{Force}}{\text{Area}}$	$\frac{MLT^{-2}}{L^2} = ML^{-1}\,T^{-2}$

Dimensional Analysis

Dimensional analysis is a useful and powerful procedure that can be used in the interconversion of units and for checking the correctness of any derived expression. Dimensional analysis is based on the fact that dimensions can be treated as algebraic quantities, that is quantities can be added or subtracted only if they have the same dimensions. This is illustrated below.

Derived Units

The units of all physical quantities can be derived from the seven base units. These units are called derived units because these can be derived from the base units algebraically by multiplication and division. Some of the common derived units are described below, and listed in Table 1.3.

1. *Area* : Area is described as,

$$\text{Area} = \text{Length} \times \text{Breadth} = (\text{Length})^2 = m^2$$

So, the SI unit of area is metre^2 (m^2)

2. *Volume* : Volume is described as,

Volume = Length × Breadth × Height

= (Length × Length × Length) (Length)^3

SI units of length is metre. So,

SI unit of volume = (metre)^3 = metre cube = m^3

3. *Density (ρ) :* The density of a substance is expressed as the mass per unit volume, *i.e.*,

$$\text{Density} = \frac{\text{Mass}}{\text{Volume}}$$

Therefore, the units of density depends upon the units of mass and volume. SI units of mass and volume are kg and dm^3 respectively. Hence,

$$\text{SI unit of density} = \frac{\text{SI unit of mass}}{\text{SI unit of volume}} = \frac{1\,\text{kg}}{1\,\text{m}^3} = 1\text{ kg m}^{-3}$$

A smaller unit of density, viz, gram per centimeter cube (g cm^{-3}) or gram per milliliter (g/mL) is commonly used.

4. *Energy :* As per definition, energy is defined as,

Energy = Force × Distance through which point of application of force moves

Therefore, the units of energy can be obtained from the unit of force and distance, viz.,

SI unit of energy = SI unit of force × SI unit of distance

= 1 N × 1 m = 1 N m

But 1 N = 1 kg m s^{-2}

Therefore, SI unit of energy = 1 kg m s^{-2} × 1 m

= 1 kg m^2 s^{-2}

This unit of energy is termed as Joule (J). Thus,

1 Joule = 1 kg m^{2}s^{-2}

The older unit of energy, calorie (cal) is related to joule through the relationship,

1 xal = 4.184 J

5. *Force :* Force is defined a

Force = Mass × Laws of Chemical Combination

and Laws of Chemical Combination = Rate of change of velocity

$$= \frac{\text{Velocity change}}{\text{Time taken}} = \frac{\text{Distance / Time}}{\text{Time}} = \frac{\text{Distance}}{\text{Time}^2}$$

So, the derived unit of Laws of Chemical Combination is given by,

SI unit of Laws of Chemical Combination

$$= \frac{\text{SI unit of distance}}{(\text{SI unit of time})^2} = \frac{\text{metre}}{(\text{second})^2} = \frac{\text{m}}{\text{s}^2} = \text{m s}^{-2}$$

Then, SI unit of force = SI unit of mass × SI unit of Laws of Chemical Combination

$$= \text{kg} \times \text{m s}^{-2} = \text{kg m s}^{-2}$$

6. Pressure (P). Pressure is defined as the force per unit area, *i.e.*,

Pressure (P) = Force (F) /Area (A)

The SI units of pressure may be obtained a follows

$$\text{SI unit of pressure} = \frac{\text{SI unit of force}}{\text{SI unit of area}} = \frac{\text{Newton (N)}}{(\text{Metre (m)})^2} = \text{N m}^{-2}$$

So, the SI unit of pressure is N m^{-2}. The unit 1 N m^{-2} is given the name of Pascal, denoted as Pa. The SI unit of pressure is 1 Pa.

The unit newton (N) may be described in terms of units of mass, length and time as follows.

Force = Mass × Laws of Chemical Combination = Mass × (Velocity/Time)

= Mass × (Distance/Time)/Time

= Mass × Distance/Time)^2

Therefore, $1 \text{ N} = 1 \text{ kg} \times 1 \text{ m}/1 \text{ s}^2$

$= 1 \text{ kg m s}^{-2}$

and $1 \text{ N m}^{-2} = 1 \text{ kg m s}^{-2} \times \text{m}^{-2}$

$= 1 \text{ kg m}^{-1} \text{ s}^{-2}$

Thus, $1 \text{ Pa} = 1 \text{ kg m}^{-1} \text{ s}^{-2}$

The derived SI unit of some common physical quantities are summarized below in Table 1.4.

Table 1.4 : Some common derived units.

Quantity	*Definition of the quantity*	*Derived SI Unit*	
		Definition	*Symbol*
Area	$(Length)^2$	m^2	
Volume	$(Length)^2$	m^3	
Density	Mass/ $(Length)^3$	kg m^{-3}	
Speed	Distance/Time	m s^{-1}	
Force	Mass × Laws of Chemical Combination		kg m s^{-2}
N (newton)			
Pressure	Force/$(Length)^2$	kg m^{-1} s^{-2}	Pa (pascal)
Energy	Force × Length	kg m^2 s^{-2}	J (joule)
Laws of Chemical	Combination	Change in speed/Time	m s^{-2}
Frequency	Cycles per second	s^{-1}	Hz (hertz)

INTERCONVERSION OF UNITS

The metric system of units is used in scientific work throughout the world. In everyday life, people in many countries such as, U. K.A., use other non-metric system of units. Therefore, it becomes frequently necessary to convert one set of units to another. This can be done in a systematic way by a method called *dimensional analysis method.*

Dimensional analysis is also called *conversion factor method. Conversion factor is equal to the ratio of the magnitudes of the same quantity in the two units.* To obtain a conversion factor, the magnitude of the physical quantity in the unit to be converted is placed in the denominator, while that in the desired unit is placed in the numerator. For example, to convert 10 minutes into seconds, one can proceed as follows.

We know, $1 \text{ min} = 60 \text{ s}$

This gives, $1 = \dfrac{60 \text{s}}{1 \text{min}}$

We can now write, $10 \text{ min} = 10 \text{ min} \times 1 \dfrac{60 \text{s}}{1 \text{min}} = 10 \times 60$

$$= \frac{\text{min} \times \text{s}}{\text{min}} = 600 \text{ s}$$

The advantage of the method is that if the equations written correctly, all the units except the desired one, get cancelled out. If the conversion factor is written incorrectly, then the desired unit is not obtained.

The conversion of units may also be done as follows. For example, to convert 10 minutes into second, one con proceed as follows.

$$10 \text{ min} = 10 \times (1 \text{ min})$$

Form the conversion tables, 1 min = 60 s

So, $10 \text{ min} = 10 \times (60 \text{ s}) = 600 \text{ s}$

To convert the units of a physical quantity involving many terms, it is found rather convenient to place the units of each physical quantity directly in the steps of the calculations along with the number, cancel out the common units and finally convert the units by using proper conversion factor.

For example, to calculate the volume of 261 g of copper having density 8.93 g/ mL one can proceed as follows :

We know, $\text{Volume of copper} = \dfrac{\text{Mass of copper}}{\text{Density of copper}}$

or, $\text{Volume of copper} = \dfrac{261\text{g}}{8.93\text{g/mL}} = \dfrac{261}{8.93}\text{mL} = 29.2\text{ mL}$

But if the volume of copper is to be expressed in litre unit, then one can proceed as follows.

$$\text{Volume of copper} = 29.2 \text{ mL}$$

$$= 29.2 \times 1 \text{ mL}$$

We know, $1000 \text{ mL} = 1 \text{ L}$

So, $1 \text{ mL} = \dfrac{1}{1000} \text{ L} = 10^{-3} \text{ L}$

Then, the volume of copper is expressed as,

$$\text{Volume of copper} = 29.2 \times 1 \text{ mL} = 29.2 \times 10^{-3} \text{ L}$$

$$= 0.0292 \text{ L}$$

Common Conversion Factors

Some common conversion factors are summarised in Table 1.5.

Table 1.5 : Some typical common conversion factors.

Physical quantity	*Unit*		*Equivalence of units*	*Conversion factor*	
	Non-SI	*SI*		*from non-SI to SI*	*form SI to non-SI*
Length	Angstrom(Å)	Metre (m)	1Å = 10^{-10}m	10^{-10} m/Å	10^{10} Å/m
	Inch (in)	"	1 inch = 0.0254m	0.0254 m.in	39.3(1/0.0254) in/m
	Centimetre(cm)	"	1 cm = 0.01m	0.01 m/cm	100 cm/m
	Mile	"	1 mile = 1609.3m	1609.3 m/mile	6.2×10^{-4} mile/m
Mass	Pound (lb)	kilogram	1 lb = 0.4536kg	0.4536 kg/lb	2.20 lb/kg
	Gram (g)	"	1 g = 10^{-3}kg	10^{-3} kg/g	10^{3} g/kg
Pressure	Atmosphere (atm)	Pascal(Pa)	1 atm = 1.0136 × 10^{5}Pa	1.0.136 × 10^{5} Pa/atm	9.86 × 10^{-6} atm/Pa
Energy	Calorie (cal)	Joule (J)	1 cal = 4.184 J	4.184 J/cal	0.239 cal/J
Power	Horse power	Watt (W)	1 hp = 745.7 W	745.7 W/hp	1.34×10^{-3}hp/W
Time	Minute (min)	Second(s)	1 min = 60 s	60 s/min	1/60 min/s
	Hour (h)	"	1 h = 3600 s	3600 s/h	1/3600 h/s

The interconversion of unit is illustrated through the following numerical problems.

REPORTING OF A MEASUREMENT

Any measurement is reported by observing the following rules.

Significant Figures

When we buy eggs, bananas, chairs, etc., we ask for an exact number and get an exact number. However, when we try to measure the height of a person with a measuring tape, it may not be possible to measure the height exactly.

The difference in the two situations arises because whereas eggs are measured by a discrete variable (there can be five eggs or six eggs but nothing in between), height is measured by a *continuous variable* (it can be 160 cm, or 161 cm or also any thing, in between). Let us suppose that the height of the person is in between 160 cm and 161 cm. Then, a measuring tape with centimetre markings can only tell that the height is greater than 160 cm and less than 161 cm, *i.e.*, the uncertainty in the value of height will be ±1 cm. If we take a scale with millimetre markings, it will be possible to measure to the nearest millimetre but not further. Here, the uncertainty in the height will be ±1 mm. Thus, *the measurement of a continuous variable can only be as precise as the choice of the measuring instrument, but no matter what we do, some uncertainty always remains.*

Measurement can be made to different degree of precision depending upon the instrument or apparatus used. It is therefore, important to include such information while reporting the results. According to the accepted convention, *a number expressing any measurement should include all digits which are certain and a last digit which is uncertain.* The total number of digits in such a number is called the number of significant figures. Thus,

"The number of significant figures in a measured quantity is equal to the number of digits whose values are known with certainty, plus the first uncertain digit"

The number of significant figures refers to the precision of a measured quantity. The concept of the number of significant figures and precision of measurement can be understood better by considering the following example.

Suppose, the height of a person has been reported in three different ways: 160 cm, 160.0cm, 160.00 cm. Although, the three ways may look equivalent, but their scientific significance is different.

The reported value of 160 cm indicates that the measurement has been made by a scale which can read only upto cm unit. The true value may lie between 159 cm and 161 cm. Thus, in the value of 160 cm, the digits 1 and 6 are certain, but 0 is uncertain. Therefore, *the measurement of 160 cm has three significant figures.*

The reported value of 160.0 cm indicates that this measurement has been made by a scale which can read upto one-tenth of a centimetre (*i.e.*, up to mm). Thus, the true value may lie between 159.9 cm and 160.1 cm. In this case, *i.e.*, in 160.0 cm, the first three digits are certain, while the fourth (*i.e.*, 0 after the decimal) digit is uncertain. Therefore, *the value 160.0 cm has four significant figures.* The scale used in this case is more precise.

The reported value of 160.00 cm indicates that this measurement has been made by a scale which can read upto one-hundredth of a centimetre (*i.e.*, 0.1 mm). Thus, the true value may lie between 159.99

cm and 160.01 cm. In this case, *i.e.*, in 160.00 cm, the first four digits are certain, while the fifth (*i.e.*, the last zero) in uncertain. Therefore, the value 160.00 cm has five significant figures. The scale used in this case is much more precise.

It is important to realise that the result of any measurement should reflect faithfully the precision of the measurement. To report more significant figures than is possible to measure in a given situation is misleading.

For example, reporting your body mass up to gram or milligram is not significant.

Rules for Determining the Number of Significant Figures

The following rules should be observed for counting the number of significant figures in a measured quantity.

Rule 1: All non-zero digits are significant. The position of the decimal point is irrelevant.

For example, the number

217 has 3 significant figures.

2.17 has 3 significant figures, and the position of decimal is not relevant.

Rule 2: A zero having digits on its right is not significant. Thus, a zero at the beginning of a number is not significant, *i.e.*, zero used for indicating the position of the decimal point, viz., 0 in 0.1 is not significant. For example,

0.314 has 3 significant figures; zero before the decimal is not significant.

0.031 has 2 significant figures; zeros on the left are not significant.

0.003 has 1 significant figures; zeros on the left of digits are not significant.

Thus in general, in any number having zeros at the beginning of the number, the zeros are not significant. For example,

0.000	1234
not significant	significant figures

Rule 3: A zero lying in between the two digits is significant. In these numbers the position of decimal is not relevant. For example,

0.304 has 3 significant figures; zero between 3 and 4 is significant.

6.023 has 2 significant figures; zero between 6 and 2 is significant.

6.0023 has 5 significant figures; both the zeros between 6 and 2 are significant.

Please note that the position of decimal is not relevant.

Rule 4: A zero having digits on its left may or may not be significant Thus, zeros after the digits may or may not be significant. For example,

250.0 has 4 significant figures; zero after the numbers is significant.

250.00 has 5 significant figures; zeros after the numbers are significant.

In exact numbers, the zeros at the end of the number also may or may not be significant. For example, 2500 may have 2, 3 or 4 significant figures, depending upon the uncertainty.

(a) If the uncertainty in 2500 is ± 100, then the second digit (*i.e.*, 5) will be uncertain. So, the number of significant figure in 2500 will be two.

(b) If the uncertainty in 2500 is ± 10, then the third digit (*i.e.*, zero) will be uncertain. So, the number of significant figures in 2500 will be three.

(c) If the uncertainty in 2500 is ± 1, then the fourth digit (*i.e.*, last zero) will be uncertain.

So, the number of significant figures in 2500 will be four.

Rule 5: In exponential numbers, the numerical portion represents the number of significant figures. Like other numbers, the zeros to the right of the decimal point are significant in exponential numbers also. For example,

(a) In the number, 2×10^3, the number of significant figures is one.

(b) In the number, 2.0×10^3, the number of significant figures is two.

(c) In the number, 2.00×10^3, the number of significant figures is three.

The rules described above are illustrated by the data presented in Table 1.6.

Table 1.6 : Illustration of rules for counting the number of Significant figures In any number.

Quantity or number	*No. of significant figures*	*Remarks*
217	3	Total number of significant digits.
2.17	3	Decimal position not relevant.
0.217	3	Zero having numbers on the right not significant.
0.00217	3	Zero having numbers on the right not significant.
7043	4	Zero in between the digits in significant.
2000	may be 2, 3 or 4	Zero having digits on the left may or may not be significant depending upon the degree of uncertainty.
2×10^3	1	Only the numerical portion represents the significant digits.
2.0×10^3	2	In exponential numbers zeros to the right of the decimal are significant.
2.00×10^3	3	-do-
2000 (exact)	4	All zeros are significant.

Exponential Numbers

This notation is used to express quantities as a multiple of 10. These numbers consists of two parts: a coefficient between 1 and 10 and a power of 10, *i.e.*, the exponential numbers are expressed as,

$$N \times 10^n$$

where, N is a number containing a single non-zero digit to the left of the decimal point.

n is an integer. It represents the power to which 10 is raised.

For example, the Avogadro's number is written as 6.02×10^{23}. Here, 6.02 is the coefficient (N), and 23 is the power of 10, (n).

The Planck's constant has a value 6.626×10^{-34} J s.

Number of significant figures in exponential numbers : In exponential numbers, the number of significant figures is given by the number of significant figures in the non-exponential number, *i.e.*, in the number N. For example, the number 5.4×10^4 has two significant figures.

It should be remembered that the number of significant figures in any measurement depends upon the precision of the measurement. For example, the Avogadro's number (N_A) may be expressed as follows in more than one way, viz.,

(a) When $N_A = 6.02 \times 10^{23}$, then there are three significant figures in this number.

(b) When $N_A = 6.023 \times 10^{23}$, then there are four significant figure in this number.

Similarly, the Planck's constant (h) can be expressed as follows.

(a) When, $h = 6.62 \times 10^{-34}$ J s, then there are three significant figures in this number.

(b) When, $h = 6.626 \times 10^{-34}$ J s, then there are four significant figures in this number.

Rounding off a Number

Rounding off a number means that the figures (digits) that are not significant are dropped. The rounding off is done to retain only the significant figures.

Rules for rounding off a number : Following rules are generally followed for rounding off a number.

Rule 1: If the digit to be dropped is greater than 5, then add 1 to the last remaining digit. For example, 62.138 will become 62.14, if rounded off to two places of decimal (to retain only four significant figures).

Rule 2: If the digit to be dropped is less than 5, then the last remaining digit is not changed. For example, 28.133, will become 28.13, (to retain only four significant figures).

Rule 3: If the digit to be dropped as 5, then the last remaining digit is left unchanged if it is even, whereas 1 is added to the last retained digit, if it is odd. For example, 1.8245 will become 1.824 (last retained digit is even, hence no change in the last retained digit) and 1.8235 will become 1.824 (last digit being odd is raised by 1).

Rule 4: If during rounding off, more than one digit is to be dropped, then these are dropped one at a time following the above rules. For example, 2.12456, if rounded off to 3 decimal places (four significant figures) would not become 2.124. Instead the rounding off done in 2 stages would give 2.125 (as follows).

1st step : 2.1246 (last digit to the dropped is greater than 5)

2nd step: 2.125 (the last digit to be dropped is greater than 5).

So, the rounded off number is 2.125.

CALCULATIONS INVOLVING SIGNIFICANT FIGURES

To express the results of an experiment, we have to often add, subtract, multiply or divide the numbers obtained in different measurements. Usually, these different numbers do not have the same precision. Therefore, it is difficult to say about the limit of precision of the final result. However, the common sense tells that when several numbers of different precision are combined, *i.e.*, added, subtracted, multiplied or divided, the final result cannot be more precise than the least precise number involved in the calculations. To report results of any measurement correctly, the following rules are found to be useful.

For Addition and Subtraction

In addition and subtraction, the final *results should be reported to the same number of decimal places as that of the term with the least number of decimal places.* The reported number should be rounded off to the desired place of decimal as per rules described above. It be noted that the total number of the significant digits in the final result would depend upon the number of decimal places required in the final result. A few examples given below shall make this point clearer.

FOR MULTIPLICATION AND DIVISION

While reporting the results of multiplication and/or division, the final results are reported to the same number of significant figures as contained in the quantity with the least number of significant figures. The presence of an exact number in the expression does not affect the number of significant figures.

ELECTROCHEMISTRY

A substance through which an electric current can pass is known as conductor. On the contrary, a substance through which no electric current can pass is known as an insulator or dielectric. An electric current in a conductor can consist of :

(a) a flow of electron only

(b) a flow of ions only, and

(c) a flow of both electrons and ions.

The first type of conductor is known as metallic conductor or electronic conductor. The second and third types of conductor are known as electrolytic (or ionic) and mixed conductors, respectively.

[I] Metallic or Electronic Conductor

When electricity passes through such type of conductor, it produces it produce no observable effect on it. It only raise the temperature of the conductor. Metals are typical example of such conductors. Certain sulphides, *e.g.*, PbS, CdS also belong to this category. In metallic conductors, there is no change in the chemical properties of the conductor.

[II] Electrolytic or Ionic Conductor

When electricity passes through such a conductor, it decomposes the conductor. Substance of this type whose solutions conduct the electricity and which are decomposed by the passage of electricity are called electrolytes. A substance whose solution does not conduct electricity is called a non-electrolyte, such as solutions of canesugar, alcohol etc.

[III] Mixed Conductor

In such a conductor, the electric current is partly electronic and partly ionic. β-Ag_2S, CuCl, γ-CuBr, metallic sodium in ammonia are examples of mixed conductors. In such a case, the ratio of electronic to ionic conductors varies with the concentration.

[IV] Electrolysis

The process of decomposition of an electrolyte by the passes of electric current through it is known as electrolysis. For bringing about electrolysis, we need a vessel known as 'electrolysis vessel'. In this vessel, an electrolytic solution is taken and two metallic rods are dipped in it. These two rods are then connected to the two terminals, (*i.e.*, + and –) of the battery. These metallic rods which lead the electric current to and form the electrolyte are called electrodes. The electrode connected with the positive end of the battery is called the *positive electrode* or *anode*, whereas the rod connected to the negative end of the battery is known as negative electrode or cathode. This complete arrangement is known as electrolytic cell. The current enters the electrolyte through the anode and leaves through the cathode.

The electrically charged atoms or radical that move to the electrodes are called ions. Ion travelling towards the cathode is known as cation and ion travelling towards anode is known as anion. The liquid surrounding the cathode is known as catholyte and that round the anode as anolyte.

[V] Laws of Electrolysis

Faraday (1834) first stated two laws on the basis of his studies on electrolysis.

1. *First Law :* According to this law, "the amount of substance (W) liberated on the electrode is directly proportional to the quantity of electricity (Q) passed."

Mathematically, $W \times Q$

If a current of C amperes is passed for t secs, then the quantity of electricity is given by current strength X time *i.e.*, c.t.

$\therefore \quad w \propto, C \cdot \square t. = Z_C t$

where, Z = proportionality constant, known as 'electrochemical equivalent.'

$$C = 1 \text{ ampere}$$

$$\text{and } t = 1 \text{ sec,}$$

$$\text{then } Z = W \text{ gms}$$

So electrochemical equivalent may be defined as, "the amount of substance (in g) liberated on the electrode on passing a current of 1 ampere for 1 second or an passing 1 coulomb of electricity."

Electrochemical equivalents of hydrogen, silver and copper are 0.00001036, 0.001118 and 0.0003292, respectively.

2. *Second Law :* According to this law, "if some quantity of electricity is passed through different electrolytes, then the amount of substance liberated on the respective electrodes will be in the ratio of their equivalent weights."

If same quantity of electricity is passed through cells containing HCl, $AgNO_3$ and $CuSO_4$, then the amounts of hydrogen, silver and copper deposited on the respective cathode will be in the ratio of this equivalent weights. Hence,

$$\frac{\text{Amount of substance liberated}}{\text{Equivalent wt. of the substance}} = \frac{\text{wt. of hydrogen}}{\text{Eq. wt. of hydrogen}}$$

$$\frac{\text{wt. of silver}}{\text{Eq. wt. of silver}} = \frac{\text{wt. of copper}}{\text{Eq. wt. of copper}}$$

or

$$\frac{\text{Wt. of hydrogen}}{1.008} = \frac{\text{Wt. of silver}}{107.88} = \frac{\text{Wt. of copper}}{31.78}$$

[IV] Faraday

It is unit of electricity and is defined as, 'the quantity of electricity (in coulombs) required to deposit or liberate 1 gm. equivalent of any substance on passing electric current.' One faraday is equivalent to 96,500 coulombs.

One the basis of faraday unit, we can find the magnitude of the elementary charge on an electron, provided Avogardro's number is known.

We know that 1 gm atom of silver requires 96,500 coulombs of electricity and the number of atoms in 1 g atom of silver is 6.023×10^{23}. So,

$$\text{Elementary change } (e) = \frac{96500}{6.023 \times 10^{23}} \text{ Coulombs}$$

$$= \frac{96500}{6.023 \times 10^{23}} \times 3 \times 10^{9} \text{ e.s.u.}$$

$$= 4.18 \times 10^{-10} \text{ e.s.u.}$$

[VII] Practical Electrical Units

The student who is acquainted with the elementary principles of electricity and magnetism is aware that electrical units are measured in various systems of units, such as electrostatic, electromagnetic and particle units. For our present purpose, it is convenient to employ practical units, a few of which we proceed to describe.

The practical unit of current is ampere which is defined as 'unvarying current which when passed through a solution of nitrate of silver in water produces silver at the rate of 0.001118 g per second.' The absolute ampere and international ampere are nearly identical.

The practical unit of quantity of electricity is coulomb, which is defined as, 'the amount of electricity conveyed in one second by a current of one international ampere.'

$$1 \text{ Coulomb} = 1/10 \text{ e.m.u.}$$

The practical unit of resistance is ohm and is defined as, 'the resistance offered to a steady current by a uniform column of mercury of length 106.300 cm at a temperature 0°C and the mass being 14.4521 g.

"The difference of potential steadily applied to a conductor, which has a resistance of 1 ohm and produces a current of 1 international ampere."

$$1 \text{ volt} = 1/300 \text{ e.m.u.}$$

For practical purpose, the international values are taken to be identical with the absolute value.

For steadily, direct currents, we have Ohm's law given by

$$E = RC$$

i.e. the emf., E of the circuit measured in volts, is equal to the product of the resistance, R in ohms and current, C in amperes. Ohm's law is not only applicable to a complete electric circuit but also to any part of the circuit.

Movement of Ions during Electrolysis

Who have already said that the flow of electricity in an aqueous solution results due to the movement of ions.

When an electric field is applied to electricity in an aqueous solution results due to the movement of ions.

When an electric field is applied to electrolytic solution, the movement of ions occurs towards respective electrodes. A filter paper strip (about 3 × 8 cm) is cut out. It is moistened with a little dilute potassium nitrate solution. Then, it is kept on a clean watch glass. A battery of 15 volts is connected across the filter paper by using a pair of crocodile clips. At the centre of filter paper strip, a suspension of yellow copper chromate in potassium nitrate is placed. When the electric current is applied across the filter paper strip, two zones are observed on the filter paper. A blue zone is seen moving towards the cathode which is due to the migration of Cu^{2+} ions. A yellow zone due to chromate ion, CrO_4^{-2}(aq) is seen moving towards the anode.

The above experiment can also be done with coloured, water-soluble ionic compounds such as potassium permanganate, cobalt chloride, nickel chloride, etc.

RESISTANCE OR CONDUCTANCE

Metallic conductors as well as electrolytes obey Ohm's law, which states that "*The strength of current (I) flowing through a conductor is directly proportional to the potential difference (E) applied across the conductor and inversely proportional to the resistance (R) of the conductor.*"

Thus, the mathematical form of Ohm's law becomes as

$$I = E/R.$$

In relation to electrolytes, the term conductance (C) is used more frequently than resistance. Conductance implies the ease with which electric current can flow through a conductor. It is therefore, defined as the reciprocal of resistance,

Conductance = 1/R.

Units: It is expressed in units of reciprocal or inverse ohms, $ohms^{-1}$ or mhos.

Resistance

The resistance R of a conductor is

(i) directly proportional to its length (l cm) and

(ii) inversely proportional to its area of cross of cross section.

Hence $$R \propto \frac{1}{a} \text{ or } R = \rho\frac{1}{a} \quad ...(1)$$

where ρ is a constant depending on the nature of the material of the conductor, and is called the "*specific resistance*" or "*resistivity.*" If l = 1 cm and a = l sq. cm, then ρ = R ohms.

Hence, specific resistance may be defined as:

"*The resistance of a uniform column of the material of the conductor having a length of 1 cm and a cross section of 1 sq. cm.*"

Units: As we know that

$$\rho = R\frac{a}{l} = \frac{\text{Ohm (cm)}^2}{\text{cm}} = \text{ohm. cm} \quad ...(2)$$

Specific Conductance

The specific conductance of a conductor is the reciprocal of specific resistance, and is generally denoted by K (Greek, Small Kappa). Equation (2) can be written as

$$R = \frac{1}{K}\frac{1}{a} \text{ ohms} \qquad \left[\therefore \rho = \frac{1}{K}\right] \quad ...(3)$$

$$K = \frac{1}{R} \times \frac{1}{a} \text{ ohms}^{-1} \text{ com}^{-1}$$

If l = 1, and a = 1 sq. cm. equation (3) becomes as

$$K = \frac{1}{R} = \rho$$

In the case of solution, the definition will be the same as of a conductor. In simple words, it is defined as:

"*The conductivity offered by a solution of length 1 cm area of unit cross section.*"

Units: We know that K = 1/ρ

Hence the unit of K will be

$$\frac{1}{\text{ohm. cm.}} = \text{ohm}^{-1} \text{ cm}^{-1} = \text{mho/cm.}$$

Equivalent Conductance

It may be defined as,

"*The conductance of a solution containing 1 gm. equivalent of an electrolyte when placed between two sufficiently large electrodes which are 1 cm apart.*"

It is denoted by λ_r, where V is the volume in c.c. containing one gram equivalent of electrolyte dissolved in it and is measured in reciprocal ohm, or mho.

Molecular Conductance

It is defined as,

"*The conductance of a solution containing 1 gm mole of the electrolyte when placed between two sufficiently large electrodes placed one cm apart.*"

It is denoted by symbol μ, and is measured in mhos,

Relation Between Specific Conductance and Equivalent Conductance

Consider a rectangular metallic vessel with opposite sides exactly 1 cm apart. If now 1 c.c. of solution is placed in this vessel, the area of the opposite faces of the cube converted by the solution will be 1 sq. cm. The measured conductance of the solution will be its specific conductançe.

If 1 c.c. of the solution is placed in the above vessel containing 1 gm. equivalent of the electrolyte, then by definition, the measured conductance will also be equal to the equivalent conductance,

$$\lambda = K$$

Now if 9c.c. of pure solvent are added so that the total volume now occupied by the solution is 10 c.c.; the (because 10 c.c. of the diluted solution still contains 1 gm. equivalent of the electrolyte). But by definition of the specific conductance, the measured conductance of the diluted solution will be ten times the specific conductance,

$$\therefore \qquad \lambda = K \times 10$$

Similarly if the solution is diluted 100 times its volume, the measured conductance will still be equivalent conductance, but will be 100 times the specific conductance.

$$\text{Thus,} \qquad \lambda_v = K \times 100$$

In general, if the solution contains 1 gm equivalent of the electrolyte dissolved in V c.c. of the solution, then

$$\lambda_r = K \times V$$

∵ Equivalent conductance = specific conductance × volume of solution in c.c. containing 1 gm. equivalent of the electrolyte.

Relation between Molecular conductance and Specific Conductance: As discussed above, a similar type of the relation also exists between molecular conductance and specific conductance *i.e.*,

$$\mu_r = K \times V$$

or Molecular conductance = [Specific conductance]

× [Volume of solution in c.c. containing 1 gm. mole of electrolyte]

Effect of Dilution

(i) *Conductance:* The conductance of s solution is due to the presence of ions in solution. Since on dilution the degree of dissociation of electrolyte in increased and more ions are produced in solution, it is thus expected that the conductance of a solution increase on dilution.

(ii) *Specific Conductance:* Specific conductance depends upon the number of ions present per c.c. of the solution. Since on dilution the degree of dissociation increases but the number of ions per c.c. decreases, it is therefore expected that the specific conductance of a solution should decrease on dilution.

(iii) *Equivalent Conductance:* The value of equivalent conductance increases on dilution. This increase is due to the fact that equivalent conductance is the product of specific conductance and the volume V of the solution containing 1 gm equivalent of the electrolyte. Since the decreasing value of specific conductance is more than compensated by the increasing value of V, the value of λ_y will increase with dilution.

(iv) *Molecular Conductance: As* discussed in equivalent conductance, the molar conductance of an electrolyte increases with dilution.

(v) The variation of equivalent conductance, where the equivalent conductance of different dilutions is plotted against $\sqrt{(c)}$, where c is the concentration of the solution in moles per litre.

It is evident that the equivalent conductance tends to become maximum as the concentration decreases or dilution increases.

However the maximum conductivity in case of strong electrolytes like KCl is reached at relatively higher concentrations (or lower dilutions) that the case of weak electrolytes like CH_3COOH.

MEASUREMENT OF E.M.F.

During the passage of electric current, the E.M.F. of an electrochemical cell is altered because:

(a) International resistance of the cell may absorbs some of the available potential difference.

(b) Chemical change may involve polarisation.

The most common producer of measuring the e.m.f. of a cell is based on Poggendroff compensation method. A working cell C usually an accumulator of constant E.M.F. (2 volts) the large then the E.M.F. of the cell to be measured is connected to the two ends of a uniform wire AB known as potentiometre wire made of platinum or platinum of high resistance. The cell S whose E.M.F. is to be measured is connected to A, with the poles in the same direction as that of the cell C and then through a galvanometer G to a sliding contact which can be moved along AB. By means of a two way key, the cell S may be replaced by a standard cell, W. First the position of sliding contact is so adjusted no current flows through the galvanometer, say at the point X, then the know that the E.M.F. of the cell, E_x would be exactly equal to the potential difference between the points A and X.

Similarly, replacing the cell S by standard cell W, the sliding contact is adjusted when no current flows through the galvanometer, say at point X′. The E.M.F. of the cell W, *i.e.*, E_w is then exactly equal to the potential difference between the two points A and X′. Therefore,

$$\frac{E_x}{E_w} = \frac{\text{Potential difference between A and X}}{\text{Potential difference between A and X}'}$$

If e is the fall in potential per unit length of the wire, then

$$\frac{E_x}{E_w} = \frac{\text{LengthAX} \times e}{\text{LengthAX}' \times e} = \frac{\text{Length AX}}{\text{Length AX}'}$$

$$\therefore \quad E_x = \frac{\text{Length AX}}{\text{Length AX}'} \times E_w$$

By means of this equation, E_x can be calculated.

ELECTRODE POTENTIAL

Nernst's theory of electrode potential. Nernst said that the metals and hydrogen have a tendency to pass into solution in the form of positive ions. This tendency is called the "solution pressure" (P), "electrolyte solution pressure" or "solution tension" of the metal and is constant at a given temperature. The metallic electrode is thus left negatively charged. So that a double electrical layer with a definite potential difference is established at the electrode surface. Since the amount of metal that goes in to the solution is minute owing to the electrostatic attraction of opposite charges, the double layer is established very rapidly. On the other hand, the ions in solution tend to be deposited on the electrodes by virtue of their osmotic pressure, P. Equilibrium is supposed to be reached when these opposing tendencies balance each other, *i.e.*, when p = P.

The establishment of a double layer can best be understood by considering the three possible cases when p > P, p < P and p = P.

(i) *When p > P :* In this case, the tendency of the ions to leave the metal will be greater than the reverse tendency. The net result will be that positive ions will enter the liquid and leave the metal negatively charged with respect to the solution. Few examples are manganese, zinc, cadmium and alkali metals, *i.e.*, metals giving the most basic oxides.

(ii) *When p = P :* In this case, the tendency for the ions to leave the metal is exactly counter balanced by the tendency to leave the solution. No double layer is thus formed and hence no potential difference exists between the metal and the solution. Such a system is known as null electrode.

(iii) *When p < P :* In this case, the tendency for the ions to leave the solution now predominates than the reverse effect. The metal in this case acquires a positive charge with respect to the solution which remains negatively charged. Metals such as copper, silver, mercury, gold etc. will be respectively charged with respect to solutions of their ions.

The solution pressure of an electrode material depends on its physical condition. Fine crystals of a metal have a higher solution pressure then do the large crystals, therefore, the former will have a more negative potential in the same solution. Solution pressure also depends upon the allotropic forms of a metal, the metastable form has a higher value. The solution pressure of a gas depends upon the pressure of the gas. The solution pressure of an elements also varies with the nature of the solvent medium containing the ions.

Nernst's expression for electrode potential—consider a metal of valency n, dipping in solution in which P, is the osmotic pressure of its ions. Let P_2 be the solution pressure of the metal and E be the actual difference of potential between the metal and the solution. A current of electricity is passed reversibly through the electrode unit 1 g ion of the metal has dissolved. Therefore, the quantity of electricity required for the dissolution of 1 g ion of the metal will be nFE coulombs. The electrical work done will be nFE volt coulombs.

Supposed the solution is now diluted, so that osmotic pressure is reduce form P_1 to $P_1 - dP_1$, the difference of potential between the metal and the solution is now changed from E to E – dE. Now in order to cause the dissolution of 1 g ion of the metal, the electrical work done or the electrical energy to be expended is (E – dE) nF volt coulombs.

The difference in electrical energy = nFE – (E – dE)nF

$$= nFdE \text{ volt coulombs.}$$

The difference in electrical energy must be equal to the osmotic work done in transferring 1 g ions of the metal from P_1 to $P_1 - dP_1$ and is equal to the product of volume and change in pressure, *i.e.*, equal to VdP, where V is the volume of the solvent containing 1 g ion of the metal.

$$\therefore \quad nFdE = VdP_1$$

If the solutions are ideal and osmotic pressures are directly proportional to the activities of the ions, then $V = RT/P_1$, so that the last equation reduces to

$$nFdE = \frac{RTdP_1}{P_1} \quad ...(1)$$

Integrating equation (45), we get

$$nFE = RT \log P_1 + \text{Integration constant}$$

When $P_1 = P_2$, then $E = 0$ ...(2)

$$\therefore \quad \text{Integration constant} = RT \log P_2$$

$$\therefore \quad nFE = RT \log P_1 - RT \log P_2$$

$$= RT \log \frac{P_1}{P_2} \quad \text{(from 46)}$$

or

$$E = \frac{RT}{nF} \log \frac{P_1}{P_2}$$

an electrode potential, represent the tendency for positive ions to pass from left to right or of negatives ions from right to left, through a cell is combined with hydrogen electrode. The potential for the electrode MIM^+ represents the tendency for the metal to pass in to solution as ions, whereas that of the electrode M^+IM is a measure of the tendency of the ions to be discharged, *i.e.*, for the ions to be reduced.

In other words, electrode potential is given a positive sign, if the electrode reaction involves reduction, *i.e.*, consumption of electrons when connected to the standard hydrogen electrode and a negative sign, if the electrode reaction involves oxidation, *i.e.*, liberation of electrons when connected to the standard hydrogen electrode taken arbitrarily as zero.

Equation (47) is known as Nernst's equation for electrode potential.

If we have two similar electrodes dipping in two different solutions, the potential difference between the two electrodes will be given by,

$$E_1 - E_2 = \frac{RT}{nF}\left(\log\frac{P_{1A}}{P_2} - \log\frac{P_{1B}}{P_2}\right) = \frac{RT}{nF}\log\frac{P_{1A}}{P_{1B}}$$

where p_1A and p^1B are the osmotic pressure of the metal ions in two solutions A and B, respectively.

As the osmotic pressure is proportional to concentration, we also have,

$$E_1 - E_2 = \frac{RT}{nF}\log\frac{c_{1A}}{C_{1B}}$$

MEASUREMENT OF ELECTRODE POTENTIAL

There is no reliable method for the determination of the absolute potential of a single electrode. The only procedure is to combine with the *reference electrode* of known P.D. and to measure the E.M.F. of the cell. If the potential of the reference electrode is arbitrarily.

Sign of Electrode Potential

The convention concerning the sign of the E.M.F. of a complete cell in conjunction with the interpretation of single electrodes potential, fixes the convention as to the sign of the electrode potentials. If we consider two cells:

(a) MIM^+ $(a_M{}^+)$, $H(a_M{}^+ = 1)$ 1 $H_2(1$ atm)

and (b) $H_2(1$ atm) 1 H^+ $(a_H{}^+ = 1)$, M^+ $(a_M{}^+)$ 1 M,

then the E.M.F. of the cell (a) will be equal and opposite to that of the cell (b). It means, therefore, that the sign of the electrode potential for the electrode MIM^+ will be equal and opposite to that written for M^+IM. The positive sign applied to taken zero, the measured E.M.F. of the cell will be equal to the potential of the other electrode on this scale. In general use, the arbitrary zero of potential is taken as that of a normal of standard hydrogen electrode.

Hydrogen Electrode

It consists of a rectangular piece of platinum foil which is connected to a piece of platinum wire sealed in a glass tube. The lower part of the glass tube is filled with mercury. The function of mercury is to keep the contact of platinum foil with the other circuit. The jacket surrounds the tube carrying the platinum foil. The jacket is closed at the top and open at the bottom. A side tube is attached to the jacket which admits hydrogen gas at normal atmospheric pressure and escapes through two holes made in the widened part of the jacket at a level midway up the platinum foil.

The hydrogen electrode is known as *standard hydrogen electrode* because both hydrogen and hydrogen ions are in their standard states. The usual convention is that the potential of the hydrogen electrode is arbitrarily assigned as zero at all temperatures. All the other potentials expressed on this basis are referred to as the potentials on the hydrogen scale.

Working: The electrode is placed in a solution of an acid. When pure hydrogen gas is passed at one atmospheric pressure, the platinum foil adsorbs a part of hydrogen and the rest comes out through the holes. Thus, there establishes an equilibrium between the hydrogen adsorbed and the hydrogen ions in solution, *i.e.*,

$$H_2 \rightleftharpoons 2H^+ + 2e^-$$

The hydrogen electrode is generally represented as:

$$Pt, H_2 \ (1 \text{ atm}) \mid H^+ \ (c = 1)$$

Whenever, the potential of any electrode say $M \mid M^+$ is to be measured experimentally, it is combined with the standard hydrogen electrode as given in the following cell.

$$M \mid M^+ \mid H^+(a = 1) \mid H_2(1 \text{ atm.}); \ Pt$$

The E.M.F. of the above cell is measured by the potentiometer and the measured E.M.F. will be equal to the required potential of the electrode on the hydrogen scale by taking potential of hydrogen electrode as zero.

Besides the above mentioned hydrogen electrodes, there are other hydrogen electrodes also. These are

(i) Pt; H_2 (1 atom) | H^+(c = 1); E = – 0.0023 volts
[normal hydrogen electrode]

(ii) Pt; H_2 (1 atm.) | N—HCl (a_H^+ = 0.81; E = 0.004) volt)

The combination used for the normal calomel electrode with hydrogen electrode may be put as

$$Pt, H2(1 atm), | H^+(a = 1), N . KCl : Hg_2 Cl_2(s) | Hg$$

Whenever calomel electrode is used in combination with hydrogen electrode it is always the positive electrode. In this, the electrons flow into the calomel electrode from the external source.

Defects in Standard Hydrogen Electrode

Although, the hydrogen electrode gives reproducible results yet it is not always convenient in practice on account of following difficulties:

(i) It is not possible to maintain the unit activity of hydrogen ions in solution.

(ii) It is not convenient to pass the gas at one atmospheric pressure uniformly.

(iii) As the platinum gets easily absorbed by the adsorption of impurities from solution and is hydrogen gas, this causes a hindrance in maintaining the equilibrium between hydrogen gas and H^+ ion in solution.

(iv) Wherever oxidising agents are present in solution, they disturb the equilibrium and which results a change in potential.

(v) This type of electrode cannot be used in the presence of reducible ions.

Calomel Electrode

As described earlier, the use of hydrogen electrode as a reference electrode is not always convenient in practice and therefore, several subsidiary electrodes have been devised. The most common of these is calomel electrode.

Construction

It consists of a glass tube having a side tube at each end. A small quantity of pure mercury is placed at the bottom of the vessel and is then covered with a paste of pure mercury, mercurous chloride and potassium chloride as shown. The apparatus is then filled with a saturated solution of KCl of known concentration through the side tube shown on the right of the vessel. The KCl solution also fills the side tube on the left which helps to make a connection through a salt bridge with the other electrode (half-cell), the potential of which is to be determined.

A platinum wire sealed into a glass tube as shown in the figure serves to make contact with mercury. When the cell is set up, it is immersed in the given solution the potential of the Calomel electrode depends upon the concentration of KCl solution employed. Its potential can be measured very accurately by connecting it to standard hydrogen electrode. The potentials of the electrode for different concentrations of KCl at 25°C are given below:

For 0.1 N KCl	E = – 0.3538 volt
For 1.0 N KCl	E = – 0.2800 volt
For saturated solution of KCl	E = – 0.2415 volt.

For greater accuracy, the electrode with 0.2N KCl is preferred due to its low temperature coefficient. In some cases saturated calomel electrode is also used because it is very easy to be set up and when it is used in conjunction with a saturated potassium chloride salt bridge, the liquid junction gets avoided.

Suppose, it is desirable to measure the electrode potential of zinc electrode immersed in N/10 $ZnSO_4$ solution. The zinc half-cell is connected with the calomel half-cell through a salt bridge of KCl. The cell so formed may be represented as:

$$Zn \mid 0.1\ N\ ZnSO_4 \parallel 0.1\ KCl,\ Hg_2Cl_2(s) \mid Hg.$$

The E.M.F. of the cell is measured by means of potentiometer. Let it be E_{obs} (say). Suppose E_{zn} be the potential of zinc electrode. As the potential of 0.1 N calomel electrode is – 0.3338 volt, hence

$$E_{obs} = E_{zn} + (-0.3338) \text{ or } E_{zn} - E_{obs} = 0.3338.$$

From the above equation it follows that E_{zn} can be calculated if E_{obs} is measured by a potentiometer.

Silver—Silver Chloride Electrode

Recently this electrode has been utilised as a reference electrode in such cases where accurate determination of chloride ions is desirable.

As it is a reversible electrode, it can be combined with such cells which contain chlorides and are free from liquid junction potentials.

It consists of a silver wire coated by a thin film of silver chloride. This deposit of AgCl may be done by using silver electrode as an anode in KCl solution by passing a current of very low density for about half an hour. The silver electrode is dipped in the solution containing chloride ions say, KCl. In order to determine its oxidation potential, it is combined with hydrogen electrode to get the complete cell as follows:

$$Ag \mid AgCl(s)\ KCl\ (aq) \mid H_2\ (1\ atm.)\ 1\ Pt.$$

The E.M.F. of the above cell is found to be – 0.2224 volts. Hence the standard potential E° is – 0.2224 volts at 25°C

Lead—Lead Sulphate Electrode

This electrode is rarely used. However, it is used in the solution containing SO_4^{2-} ions. It consists of a lead electrode dipped in lead sulphate solution. Its standard electrode potential on hydrogen scale is + 0.126. The cell may be represented by

$$Pb \mid PbSO_4(s)SO_4^{2-}$$

The cell reaction for this is

$$PbSO_4 + 2e^- \rightleftharpoons Pb + SO_4^{2-}$$

Mercury-Mercurous Sulphate Electrode

The electrode is also used in the solution containing SO_4^{2-} ions. It consists of a mercury electrode dipped in solution containing sulphate ions which is previously saturated with mercurous sulphate. This electrode may be represented by

$$Hg \mid Hg_2SO_4(s)\ SO_4^{2-}$$

APPLICATIONS OF ELECTRODE POTENTIAL MEASUREMENTS

(i) Ease of Oxidation

Electrode potential gives a quantitative idea about the relative case with which the metals are oxidised or reduced. A metal with higher positive electrode potential has greater tendency to undergo oxidation (American Convention) while one with higher – ve electrode potential has greater tendency to undergo reduction. For example, the standard oxidation potential of Li/Li^+ is 3.045 volt. So lithium is more readily oxidised to Li. On the other hand, the potential of F_2/F^- is – 2.65 volts. This shows that fluorine is very difficulty oxidised to F^- ions, or F^- ions are easily reduced to F_2.

Similarly, Fe^{2+} and Fe^{3+} posses a lower oxidation potential than that of Sn^{2+} and Sn^{4+} showing that stannous ions would reduce ferric ions whereas ferric ions would not reduce stannic ions. Thus,

$$Sn^{2+} + 2Fe^{2+} \rightarrow Sn^{4+} + 2Fe^{2+}$$

In this process, the oxidation is

$$Sn^{2+} \rightarrow Sn^{4+} + 2e^-$$

and the reduction is $2Fe^{3+} + 2e^- \rightarrow 2Fe^{2+}$

Hence the reaction can occur in the reversible cell

$$Pt \mid Sn^{2+}, Sn^{4+} \parallel Fe^{2+}\ Fe^{2+} \mid Pt$$

for the passing of two faradays. The value of E° of this cell is given by

$$E^\circ = \frac{RT}{nF} \ln K \qquad ...(1)$$

where K is the equilibrium constant. The standard oxidation potential of the two electrodes are – 0.151 and – 0.771 volt respectively. Thus,

$$E^\circ = -0.151 - (0.771) = +0.62 \text{ volt}$$

At 25°C, equation (1) becomes as

$$0.62 = \frac{0.0591}{2} \log K$$

or

$$K = \left[\frac{a_{Sn}^{4+}\ a^2{}_{Fe}{}^{2+}}{a^2{}_{Fe}{}^{2+} \times a_{Sn}{}^{2+}}\right] = 1.0 \times 10^{21}$$

The higher value of K indicates that when equilibrium is attained in ferrous-ferric and stannous stannic ions, the activities of stannous and ferric ions must be very small in comparison with those of the stannic and ferrous ions.

(ii) Replacement Tendency

Higher positive value of electrode potential indicates greater tendency of the metal to acquire the oxidised state while the higher negative value of electrode potential indicates its greater tendency to assume the reduced form. Thus, a knowledge of electrode potential gives an indication of relative replacement tendency. For example, we know at 25°,

$$E^{o}_{Zn/Zn^{2+}} = 0.761 \text{ volt}$$

and $$E^{o}_{Cu/Cu^{2+}} = -0.340 \text{ volt}$$

Therefore Zn hás much greater tendency to acquire Zn^{2+} form than copper to yield Cu^{2+} ions. In simple words, zinc will displace copper from the solution of the latter or the reaction will occur in the direction given below:

$$Zn + Cu^{2+} \rightarrow Zn^{2+} + Cu$$

(iii) Calculations of Valency in Doubtful Cases

See the "Applications of E.M.F. Measurement."

(iv) Calculation of Equilibrium Constant

We have already proved that

$$nE^{o} F = Rt \log_e K$$

or $$E^{o} = \frac{RT}{nF} \log_e K$$

Let us consider the cell

$$Zn \mid ZnSO_4 \text{ aq.} \parallel CuSO_4\text{aq} \mid Cu$$

or the passage of two faradays of electricity, the cell reaction may be represented as:

$$Zn + Cu^{2+} \rightarrow Zn^{2+} + Cu$$

The E°, the standard electrode potential, may be represented as:

$$E^{o}_{Zn', Cu} = \frac{RT}{2F} \log_e \frac{a^{2+}_{zn} a_{cu}}{a_{zn} a^{2+}_{cu}}$$

If $E_{^oZn, Zn^{3+}}$ and $E^{o}_{Cu, Cu^{2+}}$ are the standard electrode potentials of zinc and copper on hydrogen scale in their standard states, then $E^{o}_{Zn, Cu}$ gives the :ifference of potentials of zinc and copper, *i.e.*

$$E^{o}_{Zn, Cu} = E^{o}_{Zn, Zn^{2+}} - E^{o}_{Cu, Cu^{2+}}$$

$$= \frac{RT}{2F} \log_e \left(\frac{a^{2+}_{zn} a_{cu}}{a^{2+}_{cu} a_{zn}} \right)$$

But $E^{\circ}_{Zn, Zn^{2+}} = 0.761$ and $E^{\circ}_{Cu, Cu^{2+}} = -0.340$

$$\therefore\ 0.761 - (-0.340) = \frac{RT}{2F}\log_e\left(\frac{a^{2+}_{zn}a_{cu}}{a^{2+}_{cu}a_{zn}}\right)$$

or
$$1.101 = \frac{0.0591}{2}\log_e\left(\frac{a^{2+}_{zn}a_{cn}}{a^{2+}_{cu}a_{zn}}\right)$$

or
$$\left(\frac{a^{2+}_{zn}a_{cn}}{a^{2+}_{cu}a_{zn}}\right) = 1.71 \times 10^{37} = K$$

which is the value of K for the reaction occurring within the cell.

(v) Solubility Product

We have studies earlier.

(vi) Calculation of Free Energy

By knowing the E.M.F. of the cell, the free energy may be calculated as:

$$-\Delta G^{\circ} = nE^{\circ}F^{*}$$

We all know that the decrease in free energy during the working of a Galvanic cell is given by vant Hoff's isotherm as:

$$-\Delta G^{\circ} = RT\log_e K - RT\log_e\left(\frac{\text{Products}}{\text{Reactants}}\right)$$

When products and reactants are in their standard states, *i.e.* at unit activity, the above equation becomes as:

$$-\Delta G^{\circ} = RT\log_e K$$

Thus, the free energy change may be evaluated.

RATE OF ELECTRODE PROCESSES

Let us consider a metal M immersed in water to yield $[M(H_2O)_x]^+$ ion. There will be a tendency of a metal to pass into the solution as ions and then these ions in solution in turn discharge on the metal. The two processes may be represented by

$$M(H_2O)_x^{+} + 1e^{-} \underset{k_2}{\overset{k_1}{\rightleftharpoons}} M + xH_2O$$

where, k_1 and k_2 are the rate constants for the forward and backward reaction. At equilibrium, a reversible potential of the electrode will be established. The two reactions occur at equal rates at the time of equilibrium.

Glasstone, Laidler and Eyring (1940) showed that the rate of a reaction is equal to its specific rate. If a_1 is the activity of the solvated ions in solution and the activity of solid metal is unity, then the rate of the forward and reverse reaction will be k_1a_+ and k_2 respectively.

If E is the E.M.F. produced across the double layer, formed by the electrons on the metal and the ions in solution, it may be supposed that fraction α of this potential difference E facilitates the discharge of ions and the remainder (1 – α) hinders the reverse process. If z is the valency of the ions, the free energy of the discharging ions may be given by

$$\Delta G_1 = \alpha zFE$$

Similarly, the free energy of the atom of metal M which passes into the solution is given by $\Delta G_2 = (1 - \alpha)z\ FE$.

Thus,

The rate of discharge of ions from solution = $k_1a_+\ e^{azFE/RT}$ and the rate of passage of ions into solution of = $k_2e^{-(1-\alpha)FE/RT}$

At equilibrium, the two rates are equal. Thus

$$k_1a_+\ e^{azFE/RT} = k_2e^{-(1-\alpha)FE/RT}$$

or
$$e^{-zFE/RT} = a_+\frac{k_1}{k_2}$$

or
$$E = \frac{RT}{zF}\ln\frac{k_2}{k_1} - \frac{RT}{zF}\ln a_+ \qquad ...(1)$$

As k_2/k_1 is a constant at definite temperature, it means that equation (1) will take the chloride by hydrogen gas to form Ag(s) and HCl in solution. As the cell possesses no liquid junction it forms a chemical cell without transference. It is also evident from equation (7) that the E.M.F. of the cell depends upon the two factors:

(a) The activity of hydrochloric acid in solution.

(b) Pressure of hydrogen gas.

If the pressure of hydrogen is one atmosphere, ($p_{H^2} = 1$), the equation (7) reduces to

$$E = E^{o}_{Ag, AgCl} - \frac{RT}{F} \text{in } a_{H}Cl$$

Similarly, other chemical cells without transference may be easily obtained. For example, the cell is

$$H_2(P_{H2}) \mid HgSO_4(a_{H_2SO_4})(s) \mid Hg$$

It cell reaction is

$$H_2(P_{H_2}) + HgSO_4(s) \rightarrow 2Hg(l) + H_2SO_4(a_{H_2SO_4}).$$

Chemical Cells with Transference

In the chemical cell with transference the potential difference develops across the boundary between the two solutions. This potential is known as *liquid-liquid junction potential* or diffusion potential. This potential develops due to the difference in mobilities of positive and negative ions of the electrolytes.

If two solutions of different concentrations are placed in contact with one another, the more concentrated solution will have the tendency to diffuse in more dilute solution. The diffusion of each ion will be proportional to the speed of ion in the electric field. When the + ve and – ve ions of an electrolyte migrate with equal velocities, no liquid-liquid junction potential will arise.

But if the speed of – ve ion is greater than that of the positive ion, it will diffuse ahead of + ve ion into the dilute solution. With the result that dilute solution, will acquire – ve ion, the dilute solution will be + vely charged. Thus, at the junction of two solutions, a potential difference is developed up. This is termed as liquid-liquid junction potential and is represented by E_L. In other words, in the presence of liquid-liquid junction potential E_L, the E.M.F. of the cell is no longer $E_1 = E + E_2$ but instead it is

$$E = E_1 + E_2 + E_L$$

where E_1 and E_2 are two electrode potentials. An example of a chemical cell with transference is

$$\underset{E_1}{Zn \mid} ZnSO_4 \underset{E_L}{\|} AgNO_3 \underset{E_2}{\mid Ag}$$

The vertical double line in the above cell represents the liquid-liquid junction potential. The values E_1 and E_2 can be calculated as the single electrode potential of the cell, and the value of E_L is then added to the calculated total E.M.F. of the cell.

CONCENTRATION CELLS

A cell in which electrical energy is produced by the transference of a substance from a system of high concentration to one of low concentration is known as a concentration cell.

same form as the electrode potential equation. The first term on the right hand side of equation (1) is constant, and hence this gives the value of absolute single electrode potential. It is equal to the standard free energy of the conversion of the solid metal to solvated ions in solution divided by zF.

CHEMICAL CELLS

These are the cells in which the E.M.F. is due to a chemical reaction taking place within the cell. Chemical cells are of two types:

(a) Chemical Cells without Transference

In the chemical cells of this type, the liquid junction is neglected or the transference number is not taken into consideration. In these cells, one electrode is reversible to anions while the other is reversible to the anions of the electrolyte. Let us consider the cell of the following type:

$$H_2(PH_2) \mid HCl(aHCl), AgCl(s) \mid Ag$$

This is a chemical cell without transference in which the hydrogen electrode of partial pressure P_{H_2} is dipped in the solution of hydrochloric acid of activity aHCl whereas silver, silver-chloride electrode is dipped into the same hydrochloric acid. The hydrogen electrode is reversible to H^+ ions whereas silver, silver chloride electrode is reversible to Cl^- ions. The E.M.F. of the cell is equivalent to the sum of E.M.F.'s existing at the electrode solution interface. At negative hydrogen electrode, hydrogen goes into solution and oxidation occurs, *i.e.*,

$$\frac{1}{2}H_2(PH_2) \rightarrow H^+(a_H{}^+) + e^- \text{ (Oxidation)} \qquad ...(1)$$

and the E.M.F. of this half-cell is given by

$$E_1 = E^o{}_{H_2} - \frac{RT}{F}\ln\frac{a_H^+}{P_{H_2}^{1/2}} \qquad ...(2)$$

But $E°H_2 = 0$. Hence, equation (2) becomes as

$$E_1 = -\frac{RT}{F}\ln\frac{a_H^+}{P_{H^2}^{1/2}} \qquad ...(3)$$

At positive electrode the silver chloride is reduced and the reaction may be represented by

$$AgCl(s) + e^- \rightarrow Ag(s) + Cl^- (a_{Cl}^-) \text{ (Reduction)} \qquad ...(4)$$

The E.M.F. of the cell is given by

$$E_2 = E°_{Ag,\ AgCl} - \frac{RT}{F}\ln a_{\bar{Cl}} \qquad ...(5)$$

Since $a_{Ag}(s) = a_{AgCl}(s) = 0$. On adding Eqs. (1) and (4), the net cell reaction is

$$\frac{1}{2}H_2(P_{H_2}) + AgCl(s) \rightarrow Ag(s) + Cl(A_{Cl^-}) + H^+(a_H^+)$$

The total E.M.F. of this cell is given by

$$E = E_1 + E_2$$

$$= -\frac{RT}{F}\ln\frac{(a_H^+)}{P_{H_2}^{1/2}} + E°_{Ag,\ AgCl} - \frac{RT}{F}\ln a_{Cl}^-$$

$$= E°_{Ag,\ AgCl} - \frac{RT}{F}\ln\frac{(a_{H^+})(a_{Cl^-})}{P_{H_2}^{1/2}} \qquad ...(6)$$

But $(a_H^+)(a_{Cl}^-) = A_{HCl}$, hence eq. (6) becomes as

$$E = E°_{Ag,\ AgCl} - \frac{RT}{F}\ln\frac{a_{HCl}}{P_{H_2}^{1/2}} \qquad ...(7)$$

From equation (7), it follows that E.M.F. of the cell results from the reduction of silver n, then the expression of E.M.F. for 1 gm. ion of the solution may be represented as:

$$E = \frac{RT}{nF}\log_e\frac{a_1}{a_2} \qquad ...(8)$$

If the amalgams are very dilute, then they are considered to be ideal solution and, thus, the ratio of activities may be taken equal to the ratio of their concentration C_1/C_2, *i.e.*, equation (8) may take the following form

$$E = \frac{RT}{nF}\log_e\frac{C_1}{C_2} \qquad ...(9)$$

Similarly, the cell can be made by using amalgams of cadmium, lead, tin, copper and sodium.

(i) Gas Concentration Cells

In these cells, the gas electrode is generally made up of gas material at different activities and being dipped in the solution of gas ions. Generally, gases like hydrogen, chlorine and oxygen dipped in hydrogen, chlorine and oxygen ions respectively are generally employed. The potential of such type of cells depends upon the pressure of the gases and the concentration of their ions in solution. An example of this is a cell in which two hydrogen electrodes at partial pressures p_1 and p_2 are dipped in the solution of H^+ ions of activity a_{H^+}

$$\underset{p_1}{H_2(Pt)} \Big| \underset{a_{H^+}}{\text{solutin } H^+ \text{ ions}} \Big| \underset{p_2}{H_2(Pt)} \qquad ...(10)$$

When the cell operates, the oxidation will take place at negative electrode

$$\frac{1}{2}H_2(p_1) \rightleftharpoons H^+ + e \qquad ...(11)$$

and the E.M.F. of the cell may be given as

$$E_1 = E^\circ_{H_2} - \frac{RT}{nF}\log_e \frac{a_{H^+}}{p_1^{1/2}} \qquad ...(12)$$

But $E^\circ_{H_2} = 0$ and n = 1, therefore equation (12) becomes as

$$E_1 = -\frac{RT}{nF}\log_e \frac{a_{H^+}}{p_1^{1/2}} \qquad ...(13)$$

Similarly, the cell reaction taking place at positive electrode is represented as:

$$H^+ + e \rightleftharpoons \frac{1}{2}H_2(p_2) \qquad ...(14)$$

and the E.M.F. may be given as

$$E_2 = -\frac{RT}{nF}\log_e \frac{p_2^{1/2}}{a_{H^+}} \qquad ...(15)$$

The net process in this cell is

$$\frac{1}{2}H_2(p_1) = \frac{1}{2}H_2(p_2) \qquad ...(16)$$

and the total E.M.F. of this cell is $E = E_1 + E_2$.

Substituting the values of E_1 and E_2 from equations (13) and (15) in (16), we get

$$E = -\frac{RT}{F}\log_e \frac{a_{H^+}}{p_1^{1/2}} - \frac{RT}{F}\log_e \frac{p_2^{1/2}}{a_{H^+}} \quad ...(17)$$

From equation (17), it follows that the E.M.F. of this cell depends upon the partial pressure of hydrogen and is independent of the activity of the solution containing H^+ ions. In the above equation if p_2 is kept constant while p_1 is varied, the equation (17) becomes as

$$E = \frac{RT}{2F}\log_e p + \text{constant} \quad ...(18)$$

A cell in which E.M.F. depends upon the difference of concentration is called concentration cell.

Types : Concentration cells may be divided into two main classes or types:

(a) *Amalgam cells or electrode concentration cells (First Type) :* There are the cells in which the two electrodes arc of different concentrations and are dipped in the same solution of their salt, *i.e.*, the electrolyte of only one strength.

(b) *Electrolyte concentration cells (Second Type):* These are the cells in which the two electrodes are of the same material which are dipped into two solutions of same electrolyte though of different concentrations.

Of these, the second type is more important. However, here we shall discuss both these types one by one.

(a) Electrode Concentration Cells

(i) Amalgam concentration cells: In these cell, the concentration of electrode materials, usually means can be varied by utilising their solutions in mercury called amalgams. An example of this is the following cell:

$$\underset{a_1}{\text{Zn amalgam}} \left| \; ZnSO_4\,\text{solution} \; \right| \underset{a_2}{\text{Zn amalgam}}$$

In order to get different concentrations of two electrodes, zinc metal is dissolved in mercury with different activities. When the cell operate, zinc is converted into zinc ions at left hand electrode.

$$Zn(a_1) \rightarrow Zn^{2+} + 2e^- \quad ...(1)$$

and E.M.F. is given by

$$E_1 = E^\circ_{zn} - \frac{RT}{2F}\log_e \frac{a_{zn}^{2-}}{a_1} \qquad ...(2)$$

Similarly, zinc ions are converted into zinc metal at right hand electrode, *i.e.*,

$$Zn^{2+} + 2e \rightarrow Zn(a_2) \qquad ...(3)$$

and the E.M.F. is given by

$$E_2 = E^\circ_{1zn} - \frac{RT}{2F}\log_e \frac{a_2}{a_{zn}^{2+}} \qquad ...(4)$$

In this cell, the net process is

$$Zn(a_1) = Zn(a_2)$$

and the total E.M.F. is given by

$$E = E_1 + E_2 \qquad ...(5)$$

Substituting the values of E_1 and E_2 from Eqs. (2) and (4) in equation (5), we get

$$E = E^\circ_{zn} - \frac{RT}{2F}\log_e \frac{a_{zn}^{2+}}{a_1} - \frac{RT}{2F}\log_e \frac{a_2}{a_{zn}^{2+}}$$

$$= -\frac{RT}{2F}\log_e \frac{a_2}{a_1} = \frac{RT}{2F}\log_e \frac{a_1}{a_2} \qquad ...(7)$$

From equation (7), it follows that the E.M.F. of the cell is dependent upon the activity ratio of the two amalgams. If in an amalgam concentration cell the valency of the metal be

As H^+ ions are formed in hydrochloric acid solution activity a_1, then the E.M.F. of this shall may be taken as:

$$E1 = E^\circ - \frac{RT}{F}\log_e a_1 \qquad ...(8)$$

Let us now consider the cell with different activities of HCl. This is represented by

$$H_2(1\ atm) \mid HCl(a_2)AgCl(s) \mid Ag$$

As proved earlier, the net reaction occurring in this cell is

$$\frac{1}{2}H_2(1\ atm) + AgCl(s) \rightarrow HCl(a_2) + Ag(s) \qquad ...(9)$$

The E.M.F. (E_2) for this cell is

$$E_2 = E^\circ - \frac{RT}{F}\log_e a_2 \qquad ...(10)$$

Now connect the two cells containing hydrochloric acid of activities a_1 and a_2 with E.M.F.'s E_1 and E_2 in opposition. The resulting cell will be

H_2(1 atm) | HCl(a1), AgCl(s) | Ag | AgCl(s), HCl(a_2) | H_2(1 atm.)

The E.M.F. of the above cell is $E_1 - E_2$ and the net reaction may be obtained by subtracting Eq. (9) from (8):

$$HCl(a_2) = HCl(a_1) \qquad ...(11)$$

Now the electromotive force E of the cell is

$$E = E_1 - E_2$$

$$= \left(E^\circ - \frac{RT}{F}\log_e a_1\right) - \left(E^\circ - \frac{RT}{F}\log_e a_2\right)$$

$$= \frac{RT}{F}\log_e \frac{a_2}{a_1} \qquad ...(12)$$

As $a_2 > a_1$, there occurs a transfer of electrolyte from concentrated to dilute solution which follows from equation (12).

Now we know $a = m^2\gamma^2$.

where m is the molality and γ is the activity coefficient. Then,

$$E = \frac{RT}{F}\log_e \frac{m_2^2\gamma_2^2}{m_1^2\gamma_1^2} = \frac{2RT}{F}\log_e \frac{m_2\gamma_2}{m_1\gamma_1} \qquad ...(13)$$

The activities may be expressed in terms of concentrations, *i.e.*, $a = c^2f^2$. Then, equation (12) becomes as

$$E = \frac{RT}{F}\log_e \frac{c_2^2 f_2^2}{c_1^2 f_1^2} = \frac{2RT}{F}\log_e \frac{c_2 f_2}{c_1 f_1} \qquad ...(14)$$

For dilute solution $f_2 = f_1 = 1$, then equation (14) becomes as:

$$E = \frac{2RT}{F}\log_e \frac{c_2}{c_1} \qquad ...(15)$$

From equation (9), it follows that the E.M.F. of the cell without transference depends upon the concentration of the solution in both sides. Equation (12) may also be expressed in the form given below:

$$E = \pm \frac{v}{v \pm z \pm} \frac{RT}{F} \log_e \frac{a_2}{a_1}$$

where z_+ and z_- are valencies positive and negative ions with respect to electrodes with which these are reversible and v_+ and v_- are the number of positive and negative ions, and v is total number of ions.

It follows that the plot of E.M.F. of the cell against $\log_e p$ gives a straight line. Hence in hydrogen concentration cell if one hydrogen electrode is kept at constant pressure, the E.M.F. possesses a linear exponential relationship with the pressure of the gas.

Similarly, other gas cells can be constructed. Few examples are as follows:

(i) In the case of oxygen cell, the reaction occurring may be put as

$$O_2 + 2H_2O + 4e^- \rightleftharpoons 4OH^-$$

In thick cell, the transfer of one mole of oxygen from one electrode to other requires the passage of 4 Faradays. The E.M.F. of such a cell at two oxygen electrodes at different pressures may be given by

$$E = \frac{RT}{4F} \ln \frac{a_2}{a_1} \qquad ...(16)$$

If the gas behaves ideally, equation (16), becomes as

$$E = \frac{RT}{4F} \ln \frac{p_2}{p_1} \qquad ...(17)$$

The sign of the E.M.F. of this cell is just opposite to that the corresponding hydrogen cell as represented by eq. (17).

(ii) It is possible to prepare cells where the electrodes are of different gases. The hydrogen chlorine cell may be shown as

$$H_2(Pt) \mid H^+ \text{ solution} \mid Cl^- \text{ solution} \mid Cl_2(Pt)$$

The Grove cell, *i.e.*, the hydrogen oxygen is also an example of this type:

$$H_2(Pt) \left| \begin{array}{c} \text{Water} \\ \text{containing acid, alkali} \\ \text{or salt} \end{array} \right| O_2(Pt)$$

(b) Electrolyte Concentration Cell

This type of concentration cells may be further divided into two sub-types:

(i) *Concentration cells without transport or transference:* In these types of cells, there is not direct transfer of electrolyte from one solution to another. The transfer occurs due to the result of chemical reactions. In general,

"*A concentration cell without transference can be designed if two simple cells with electrodes reversible with respect to each of the ions constituting the electrolyte are oppositely combined.*"

Let us consider one cell having the electrodes reversible with respect to H^+ and Cl^- as:

$$H_2(1 \text{ atom}) \mid HCl(a_1)\ AgCl(s) \mid Ag$$

When the cell starts functioning, the hydrogen at left hand electrode dissolves to form hydrogen ions whereas from silver chloride, silver is deposited at the right hand electrode as represented by

$$\frac{1}{2}H_2(1 \text{ atm}) \rightarrow H^+ + 1e \qquad \text{(left hand electrode)}$$

$$AgCl(s) + 1e \rightarrow Ag + Cl^- \qquad \text{(right hand electrode)}$$

The net cell reaction may be obtained by adding the above two half-reactions.

$$\frac{1}{2}H_2(1 \text{ atm}) + AgCl(s) \rightarrow HCl(a_1) + Ag(s)$$

But $$E_1 = E° - \frac{RT}{F}\log_e a_1{}^{t_a}$$

and $$E_2 = E° - \frac{RT}{F}\log_e a_2{}^{t_a}$$

$$E = \left(E° - \frac{RT}{F}\log_e a_1{}^{t_a}\right) - \left(E° - \frac{RT}{F}\log_e a_1{}^{t_a}\right)$$

or $$E = -\frac{RT}{F}\log_e a_1{}^{t_a} + \frac{RT}{F}\log_e a_2{}^{t_a}$$

or $$E = -\frac{RT}{F}\log_e \frac{a_1{}^{t_a}}{a_2{}^{t_a}}$$

$$E = -t_a\frac{RT}{F}\log_e \frac{a_1}{a_2} - t_a\frac{RT}{F}\log_e \frac{a_2}{a_1}$$

We know $a = m^2\gamma^2$ where m is the molality and γ is the activity coefficient. Then,

$$E = t_a \frac{RT}{F} \log e_e \frac{m_2^2 \gamma_2^2}{m_1^2 \gamma_1^2}$$

$$= 2t_a \frac{RT}{F} \log_e \frac{a_2}{a_1}$$

In the case of concentration cell with transport, the E.M.F. in terms of valencies may be put as

$$E = t_a \frac{v}{v \pm z \pm} \frac{RT}{E} \log_e \frac{a_2}{a_1}$$

where v is total no. of ions in solution, in z denotes the valency of ions. In the cell under consideration the total number of ions, *i.e.*, v = 2, v± = , then the E.M.F. may be given as

$$E = 2t_a \frac{RT}{F} \log_e \frac{a_2}{a_1}.$$

CHEMICAL EQUILIBRIUM

A reaction is said to be reversible if the composition of the reaction mixture on the approach of equilibrium at a given temperature is the same irrespective of the initial state of the system, *i.e.*, irrespective of the fact whether we start with the reactants or the products. A reversible reaction never tends to completion.

Some examples of reversible reactions are as follows:

$$Fe + 4H_2O \rightleftarrows Fe_3O_4 + 4H_2$$

$$N_2 + O_2 \rightleftarrows 2NO$$

$$H_2 + I_2 \rightleftarrows 2HI$$

The extent of reversibility of a reaction is governed by the nature of the reaction and the experimental conditions of temperature, concentrations of substances and pressure on the system, if gases are involved. Most of the reaction that we know are reversible in nature.

IRREVERSIBLE REACTIONS

A reaction in which the product do not interact to form the original reactants is called an irreversible reaction. Some examples of irreversible reactions are as follows:

$$2KClO_3 \rightarrow KCl + 3O_2$$

$$NaCl + AgNO_3 \rightarrow AgCl + NaNO_3$$

Irreversible reactions proceed to completion in one direction. Irreversible reactions have their equilibrium positions almost on one extreme.

Most of the precipitation reactions are irreversible.

CHEMICAL EQUILIBRIUM

A chemical equilibrium can be defined as a *dynamic state at which the rate of forward and backward reactions are the same and at that instant all the reactants and products are present.* When chemical equilibrium is reached both the forward and backward reactions continue but at the same ate so that the concentrations of the components remain unchanged with time. Chemical equilibrium may be homogeneous or heterogeneous depending upon whether the reaction takes place :

(i) In one phase or

(ii) In two or ore phases.

Examples:

(i) When stream is passed over iron in a closed vessel, it will attain the state of equilibrium and the reactants and products will have a definite concentration at a given temperature.

$$2FE + 4H_2O \rightleftarrows Fe_3O_4 + 4H_2$$

(ii) If a mixture of hydrogen and iodine vapour is passed over platinum heated to 450°C in a closed vessel, the formation of hydrogen iodide and its decomposition into hydrogen and iodine takes place with equal and opposite rate and a state of equilibrium is reached.

$$H_2 + I_2 \rightleftarrows 2HI.$$

Dynamic Nature of Chemical Equilibrium

A reversible reaction must attain the equilibrium state. For instance, if hydrogen and iodine are taken in a reaction vessel, only toward reaction takes place as initially there is no hydrogen iodide. As soon as hydrogen iodide is formed, the reverse reaction starts. As time passes, the concentrations of hydrogen and iodine start decreasing and, therefore, the rate of forward reaction decreases.

However, the rate of reverse reaction increases. As time passes, an equilibrium state may be reached when the rates of toward and reverse reactions become equal. At this stage no further change in consecrations occurs.

Similar results would be obtained if we start with HI. At equilibrium, concentrations of HI, H_2 and I_2 become constant at a certain temperature. This condition is known as chemical equilibrium.

Recognition of Attainment of Equilibrium

It is possible to recognize the attainment of equilibrium by constancy of certain observable properties. For example, the attainment of equilibrium is recognised in the evaporation of water at a given temperature by observing the pressure had become constant. Another example is the dissociation of calcium carbonate in which the attainment of equilibrium is recognised by observing that pressure of carbon dioxide had become constant. In both the examples, it becomes possible to recognise the attainment of equilibrium by observing constancy of pressure.

In some cases, the measurement of concentration of one or more of the reactants or products can be used to recognise the attainment of equilibrium. In order to understand this, we will consider the reaction between hydrogen gas and iodine vapour.

$$H_2(g) + I_2(g) \rightleftarrows 2HI(g)$$

Suppose one mole of hydrogen gas and one mole of iodine vapour are mixed in a flask of 1 litre capacity. Now suppose this flask is allowed to stand in a boiling bath (448°C) for several hours. If the reaction goes to completion, one should expect the formation of two moles of hydrogen iodide. But it is observed that the reaction comes to a stop when only 1.56 moles of hydrogen iodide get formed. This reveals that only 0.78 mole of each of hydrogen and iodide have reacted. Thus, it is evident that 0.22 mole of hydrogen and 0.22 mole of iodine are still present as such without reacting. If the mixture is kept at the same temperature even for a couple of days, there will occur no change in the concentration of either two products or any of the reactants. In this reaction, the attainment of equilibrium is recognised by observing constancy of concentration.

A well known equilibrium reaction between $N_2O_4(g)$ and $NO_2(g)$ is considered,

$$\underset{\text{Almost colourless gas}}{N_2O_4(g)} \rightleftarrows \underset{\text{Deep reddish gas}}{2NO_2}$$

In this reaction, the colour of the gaseous mixture becomes constant at equilibrium.

From the above discussion, it can be concluded that the attainment of equilibrium can be recognised by observing constancy of properties such as pressure, concentration, density of colour whichever may be found to be suitable in a given case.

Characteristics of Chemical Equilibrium

Let us now consider certain important characteristics of chemical equilibrium. These are:

1. Chemical equilibrium, at a given temperature, can be characterised by constancy of certain observable properties like pressure, concentration, density or colour.
2. Chemical equilibrium can be approached from either direction, let us take an example to illustrate this important characteristic of chemical equilibrium. We know that hydrogen gas and iodine vapour react at elevated temperature yielding hydrogen iodide gas, according to the equation.

   $$H_2(g) + I_2(g) \rightarrow 2HI(g)$$

   We know that hydrogen iodide is not a very stable compound. This on heating dissociates into hydrogen and iodine.

   $$2HI(g) \rightarrow H_2(g) + I_2(g)$$

   It means that the hydrogen-iodine reaction is reversible and can be represented as

   $$H_2(g) + I_2(g) \rightleftarrows 2HI(g)$$

   When 1 mole of H_2 is allow to mix with 1 mole of iodine in a closed vessel of one litre capacity and the vessel is allowed to stand for a long time at 448°C (the temperature of boiling sulphur), ultimately, only 1.56 moles of HI, gets formed. There occurs of further increase in the amount of HI even if the reaction is permitted to continue for several days at the same temperature of 448°C. This reveals that only 0.78 mole of each of H_2 and I_2 has reacted. The composition of the reaction mixture at equilibrium therefore, would be 1.56 moles of Hi, 0.22 mole of H_2 and 0.22 mole of I_2.

Now the reaction is carried out by taking 2 moles of Hi i the same vessel and keeping it again in the both of boiling sulphur at 448°C. After about the same time as before, it will be found that only 0.44 mole of HI gets dissociated into H_2 and I_2. The composition of the reaction mixture, therefore, would be 1.56 moles of HI, 0.22 mole of H_2 and 0.22 mole of I_2 as before.

This reveals that when equilibrium gets attained at a given temperature each reactant and each product is having a fixed concentration and this is independent of the fact whether the reaction is started with the reactants or with the products. Thus, equilibrium can be attained from either side.

3. A catalyst can hasten the approach of equilibrium but fails to alter the state of equilibrium. In other words, it means that the relative concentrations of the products and the reactants remain the same irrespective of the presence or absence of a catalyst. For example,

   $2SO_2(g) + O_2(g) \rightleftarrows 2SO_3(g)$

   This reaction is allowed to take place, generally, at 500°C. The equilibrium in this case would be only attained after a long time. However, if a catalyst (platium or vanadium pentoxide) is add to the system, the equilibrium would be attained fairly rapidly but the equilibrium state is not altered. In other words, the relative amounts of SO_2 (the product) and SO_2 and O_2 (the reactants) at equilibrium would remain the same irrespective of the fact whether a catalyst is added or not in the reaction. Hence the yield of SO_2 would not be improved in any way by using the catalyst.

4. Chemical equilibrium is dynamic in nature because it involves two opposing reactions. One of these reactions takes place from the reactants towards the products and is known as the forward reaction. The other takes place from the products towards the reactants ad is known as the reverse reaction. When equilibrium is attained, there occurs no further change in the concentrations of the products or the reactants. This gives the idea that the reaction has stopped. But this is not the case. Actually, the opposing reactions, the forward reaction ad the reverse reaction,

are taking place simultaneously at equal rates. Hence one reaction completely undoes the effect of the other. In other words, as much of the products get formed in a given time as change back into the reactants i the same time. Hence the chemical equilibrium is dynamic and not static.

EXPERIMENTAL PROOF FOR DYAMICAL EQUILIBRIUM

Now, it has been possible to establish the dynamic nature of chemical equilibrium experimentally in several cases. This has been demonstrated in the case involving hydrogen-iodine-hydrogen iodide reaction.

$$H_2(g) + I_2(g) \rightleftarrows 2HI(g)$$

The equilibrium of this reaction is attained at 448°C. As soon as this equilibrium is attained, a radioactive isotope of iodine is introduced into the reaction mixture. After sometime, it will be found that HI contains radioactive iodine whereas the relative amounts of HI, H_2 and I_2 do not get changed. This reveals that although no change is taking place in the relative amounts of the reactants and products, yet chemical reaction is taking place from left to right. But there occurs no increase i the amount of Hi. Therefore, it means that the reaction is taking place from left to right as well at the same rate. Thus, it means that although the system has been in equilibrium, yet the two opposing reactions have been proceeding and the equilibrium conditions are being maintained by a dynamic balance between the two reactions.

LIMITATIONS OF THE EQUATION FOR CHEMICAL EQUILIBRIUM

The equation for representing chemical equilibrium falls to reveal how far the reaction would proceed from left to right or right to left before the equilibrium is attained. Suppose the following equilibrium is considered:

$$N_2O_4(g) \rightleftarrows 2NO_2(g).$$

The above equation does not reveal what fraction of N_2O_4 gets dissociated into NO_2. If the temperature of the reaction is raised by 10°C, there occurs a considerable increase in the dissociation of N_2O_4 but even then the equation for the reaction remains the same.

The equation used for representing the chemical equilibrium fails to reveal how long it takes for a reaction to attain equilibrium. This can be understood by considering the equilibrium of water vapour with hydrogen and oxygen

$$H_2O(g) \rightleftarrows H_2(g) + \frac{1}{2}O_2(g)$$

Water is a highly stable compound. Its degree of dissociation is appreciably small. Even at 2000°C, its dissociation is mainly 0.6%. Thus, the equilibrium is lying towards much to the left. However, it has been verified experimentally that a mixture of hydrogen and oxygen remains as such for a sufficient long time without undergoing reaction to form water and establish the above equilibrium. The equation for the equilibrium fails to reveal that it requires so long to the equilibrium.

If an electric arc is struck in the hydrogen-oxygen mixture, the reaction starts instantaneously and the equilibrium is attained within minute or so. The equation of the chemical equilibrium fails to reveal the rate at which a chemical system comes to equilibrium and the conditions under which the equilibrium can be attained rapidly.

LAW OF MASS ACTION

A qualitative relation connecting the concentrations of the reactants and products with the amount of chemical change produced known as the low of mass action was proposed in 1867 by Guldbeg and Waage. This law can be stated as follows:

"The rate at which a substance reacts a proportional to the active mass and the rate of a chemical reaction is proportional to the product of active masses of the reactants."

By the term active mass, it means the *molecular concentration.* In gases and solutions, it means the umber of gram molecules (*i.e.*, moles) present per litre. In the case of solids or pure liquids the active mass may be regarded as constant, *i.e.*, for sake of simplicity, it is taken as unity.

Active mass of a substance is expressed by enclosing the symbol or the formula of a substance i square brackets *i.e.*, [A] or by the symbol C at the base of which the symbol of the substance is placed, *i.e.*, C_A

Let us consider the following reversible reaction taking place at constant temperature:

$$A + B \rightleftarrows C + D$$

According to the law of mass action, the velocity of forward reaction: v_1 is given by

$$v_1 \propto C_A + C_B$$

$$v_1 \propto k_1C_A + C_B \quad ...(1)$$

where C_A and C_B are the molecular concentrations of A and B and k_1 is a constant which may be called the *velocity constant of the forward reaction.* The concentrations of A and B go on decreasing with the progress of the reaction and hence the velocity of the forward reaction gradually decreases with time as shown graphically in Fig. 1. Since the reaction is reversible, the products as soon as they accumulate, start reacting to reform the reactants. The velocity of backward reaction is given by

$$v_2 = k_2 + C_C + C_D \quad ...(2)$$

where C_C ad C_D are the molecular concentrations of C and D and k_2 is the velocity constant of the backward reaction. As the concentrations of the products gradually increase, the velocity of the backward reaction will also steadily increase, as shown in Fig. 4.1.

Since the commencement of the reaction (zero time), the concentrations of the products gradually increase, the velocity of the backward reaction will also steadily increase. Ultimately a stage called *dynamic equilibrium* is reached when the velocity of toward reaction becomes equal to that of backward reaction, *i.e.*,

$$v_1 = v_2 \quad ...(3)$$

At this stage the number of products molecules produced per second in the forward reaction would be equal to the number of product molecules which would disappear in the backward reaction.

Equilibrium Constant

On substituting Eqs. (1) and (2) in Eq. (3), we get

$$k_1CAcB = k_2C_CC_D$$

or

$$\frac{k_1}{k_2} = K_C = \frac{C_CC_D}{C_AC_B} \quad ...(4)$$

where K_C is a constant, called the equilibrium constants. It is defined as the ratio of velocity constants of two opposing reactions.

Eq. (4) is known as the equilibrium law equation.

When the reaction is of the type $2A \rightleftarrows C + D$, one may write

$$2A \rightleftarrows C + D$$

Hence the equilibrium constant is given by

$$k_C = \frac{C_C \times C_D}{C_A C_A} = \frac{C_C \times C_D}{C_A^2}$$

A general equation for a reversible reaction can be written as

$$aA + bB + ... \rightarrow cC + dD + ...$$

Hence the equilibrium constant is given by

$$k_C = \frac{C_C^c \times C_D^d \times ...}{C_A^a \times C_B^b \times ...} \qquad ...(5)$$

From Eq. (5), it follows that the law of mass action may be stated in another way

"At a give temperature, the rate of a reaction is proportional to the molar concentration of arch reacting species raised to the power of the number of its molecules taking part in the reaction as represented in the stoichiometic equation describing the reaction."

In Eq. (5), when the concentrations are expressed in moles per litre, the equilibrium concentration is written as K_C.

In the case of gaseous reactions, the concentrations may be put proportional to the partial pressures. Hence the gaseous reactions, Eq. (5) may be put in the following form:

$$K_P = \frac{P_C^c \times P_C^d \times ...}{P_A^a \times P_B^b \times ...} \qquad ...(6)$$

where K_P is another equilibrium constant. The partial pressures are generally expressed in atmospheres.

Relationship between K_P and K_C. For an ideal gas,

$$P_i V = nRT$$

cr

$$P_i = n\frac{RT}{V} = C_i RT \qquad ...(7)$$

where C is the concentration. On substituting the concentrations of various species from Eq. (7), in Eq. (6), we get

$$K_P = \frac{(C_C RT)^c \times (C_D RT)^d \times ...}{(C_A RT)^a \times (C_B RT)^b \times ...}$$

$$K_p = \frac{C_C^c \times C_D^d \times ...}{C_A^a \times C_B^b \times ...} \times (RT)^{(c+d)-(a+b)}$$

$$K_p = K_C \times RT^{\Delta n}$$

where $\Delta n = (c + d + ...) - (a + b + ...)$

Equilibrium Constant : The equilibrium constant is defined as simply the ratio of the velocity constants of two opposing reactions.

OR

The equilibrium constant is the product of the concentrations of the reacting species, each concentration raised to the power which is the stoichiometric coefficient of the substance in the chemical equation.

In general, for the reaction represented by the equation,

$$aA + bB + cC \rightleftarrows lL + mM + nN$$

the equilibrium constant is give by

$$K_C = \frac{[K_1]}{[K_2]} = \frac{[L]^l[M]^m[N]^n}{[A]^a[B]^b[C]^c}$$

The equilibrium constant for the reaction $2SO_2 + O_2 \rightleftarrows 2SO_3$ is as follows:

$$K_c = \frac{[SO_3]^2}{[SO_2]^2[O_2]} = \frac{(\text{litre / mol})}{(\text{litre / mol})(\text{litre / mol})^2}$$

It means that the equilibrium constant for this reaction is constant.

Factors Influencing Equilibrium Constant

The various characteristics or the reactors that influence the equilibrium constants are as follows :

1. *Change in Temperature :* The equilibrium constant has a definite value for every chemical reaction at a given temperature. The value of equilibrium constant varies with temperature. It has also the same value (at a give temperature) irrespective of the pressure on the system or the concentrations of the reactants and products.
2. *The Mode of Expressing the Reaction :* If a reaction is represented in two opposite ways, the value of equilibrium constant in each case would be the reciprocal of the equilibrium constant i th other case For example,

$$A + B \rightleftarrows C + D, \quad K = \frac{C_c \times C_d}{C_A \times C_B}$$

$$C + D \rightleftarrows A + B, \quad K_1 = \frac{C_A \times C_B}{C_C \times C_D} = \frac{1}{K}$$

It means that if the equilibrium constant is large in one direction, it will be smaller in the other direction.

3. *Change of Concentration or Partial Pressure Units :* The units of equilibrium constants K_p and K_C are

$$K_P = (\text{Atm})^{\Delta n},$$

$$K_C = (\text{Mol dm}^{-3})^{\Delta n}$$

For reactions in which Δn is zero, *i.e.*, the number of moles of products is same as that of reactants, K has no units. For reactions in which Δn is not zero, any change in units expressing concentration or partial presents of the involved substances will alter the equilibrium constant.

4. *Representation of Stoichiometric Equation :* The value of equilibrium constant depends upon the stoichiometric equation representing the reversible reaction. If it is expressed in different ways, the value of K is affected. The formation of SO_3 from SO_2 and O_2 may be represented by either of the following stoichiometric equations.

$$2SO_2 + O_2 \rightleftarrows 2SO_3 \quad \text{...(i)}$$

$$SO_2 + \frac{1}{2} \rightleftarrows SO_3 \quad \text{...(ii)}$$

K_1 from reaction (i),

$$K_1 = \frac{C_{SO_3^2}}{C_{SO_2} \times C_{O2}} \quad \text{...(iii)}$$

K_2 from reaction (ii),

$$K_2 = \frac{C_{SO_3}}{C_{SO_2} \times C_{O2}^{1/2}} \quad \text{...(iv)}$$

From (iii) and (iv), we have

$$K_2 = \sqrt{K_1}$$

Significance of Equilibrium Constant

The equilibrium constant is regarded as a convenient measure of the extent to which a reaction goes before equilibrium is established. In order to illustrate the statement, we consider the following reaction:

$$2H_2O(g) \rightleftarrows 2H_2(g) + O_2(g);$$

$$K = 1.35 \times 10^{-11} \text{ mol l}^{-1} \text{ at } 1073 \text{ K}$$

As the value of K is very small, it means that there will be very small quantities of H_2 and O_2 before equilibrium is established. Thus, the reverse reaction is almost complete. Due to this reaction, a mixture of H_2 and O_2 is dangerously explosive.

Let us now consider another reaction

$$2O_3(g) \rightleftarrows 3O_2(g);$$

$$K = 10^{55} \text{ mol l}^{-1} \text{ at } 298 \text{ K.}$$

because K is very large, the reaction mixture at equilibrium will contain a high proportion of oxygen and reaction will be nearly complete at equilibrium, *i.e.*, O_2 is more stable than O_3.

DE DONDERS CONCEPT OF DEGREE OF ADVANCEMENT OF A REACTION

In order to understand this concept, *i.e.*, the progress of a reaction, we shall consider the hypothetical reaction of the following type which is proceeding in a closed vessel.

$$aA + bB \rightleftarrows cC + dD.$$

As this reaction is being carried out in a closed vessel, it means that the total mass of the system, evidently, will remain constant although the number of moles of each reactant and product will go on changing with time. If ζ represents the progress or degree of advancement of the reaction then $d\zeta$ will represent the small increase in the extent of the reaction that takes place i any given interval of time. Due to this, the amount of A decreases by a $d\zeta$ and that of B by b $d\zeta$ while the amount of C increases by cd ζ and that of D increased by d $d\zeta$. Mathematically, these changes may be represented as follows:

$$dn_A = -ad\zeta \qquad dn_B = -bd\zeta$$

$$dn_C = -cd\zeta \qquad dn_D = -dd\zeta$$

If the temperature and pressure of the system are assumed to be constant, the change in Gibbs free energy for this small progress of this reaction will be as follows:

$$(dG)_{T,P} = \sum_{1} \mu_i dn_i$$

$$= (c\mu_C + d\mu_D - a\mu_A - b\mu_B)d\zeta$$

or $$\left(\frac{\partial G}{\partial z}\right)_{T,P} = (c\mu_C + d\mu_D - a\mu_A - b\mu_B)$$

$$= \sum_I v_i\mu_i$$

where v_i is the difference between the stoichiometric coefficients of the products and those of the reactants, *i.e.*, (c + d) – (a + b).

De Donder introduced the term $\left(\frac{\partial G}{\partial \zeta}\right)_{T,P}$ which may be defined as the change of *Gibbs free energy per unit change in the extent of the reaction*. Three cases may arise:

(i) If the value of this term is negative, the reaction will proceed from left to right.

(ii) If the value of this term is positive, the reaction will proceed from right to left.

(iii) If the value of this term is zero, th system will be in a state of equilibrium.

DERIVATION OF LAW OF MASS ACTION FROM CHEMICAL POTENTIAL

Consider a reaction

$$pA + qB + ... \rightleftarrows rC + mD + ...$$

Suppose the concentrations of the reactions A, B ... at the start of reaction are C_A, C_B ... respectively. Let the concentrations of the products C and C, ... at the end of reaction be C_C and C_D ... respectively. The free energies per mole (*i.e.*, chemical potentials) of the various substances at temperature T will be denoted by μ_A, μ_B, μ_C, μ_D .... respectively.

We know that the free energy of a system consist in of several components at constant temperature and pressure is give by

$$G = n_1\mu_1 + n_2\mu_2 + ... \qquad ...(1)$$

where n_1, n_2 ... are the number of moles of the various components and m_1, m_2, ... are their respective chemical potentials. Hence one may write

Free energy of products = $r\mu_C + m\mu_D$

Free energy of reactants = $p\mu_A + q\mu_B$

$\therefore$ $\Delta G = (r\mu_C + m\mu_D) - (p\mu_A + q\mu_B)$ ...(2)

The chemical potential of a substance in any state is give by the equation

$$\mu = \mu^o + RT \ln a$$

where μ^o is chemical potential in the standard state of unit activity and μ the chemical potential at activity a.

Replacing the values of chemical potentials of the products and the reactants in equation (2), we get

$$\Delta G = [r(\mu_C^o + RT \ln a_C) + m(m_D^o + RT \ln a_D)] + ...$$
$$-[p(\mu_A^o + RT \ln a_A) + q(\mu_B^o + RT \ln a_B) + ...]$$

Rearranging,

$$\Delta G = [(r\mu_C^o + m\mu_D^o + ...) - (p\mu_A^o + q\mu_B^o + ...)]$$
$$+ RT \ln \frac{(a_C)^r (a_D)^m ...}{(a_A)^p (a_B)^q ...} \quad ...(4)$$

The μ^o terms refer to standard states of unit activity of the substance concerned. The first term in square brackets on the right side of equation (4), therefore, represents the increase in free energy of the reactants when these at unit activity react to form products also at the unit activity. If this energy increase is represented by ΔG^o, equation (4) may be written as

$$\Delta G = \Delta G^o + RT \ln \frac{(a_C)^r (a_D)^m ...}{(a_A)^p (a_B)^q ...} \quad ...(5)$$

$$= \Delta G^o + RT \ln x \quad ...(6)$$

where x stands for reaction quotients of activities of the products and reactants, viz.,

$$x = \frac{(a_C)^r (a_D)^m ...}{(a_A)^p (a_B)^q ...}$$

Equation (6) is commonly known as the reaction isotherm.

If the above reaction is in equilibrium state, then ΔG will be zero. Hence from equation (5),

$$\Delta G^o = -RT \ln \left(\frac{(a_C)^r (a_D)^m}{(a_A)^p (a_B)^q} \right) \quad ...(7)$$

Since ΔG^o represents standard free energy change, *i.e.*, when the reactants as well as products are in their standard states of unit activity it must be constant at a given temperature. Consequently the expression on the right side of equation (7) must be constant as well. Since R is a gas constant, it follows, therefore, that if the temperature T is constant, then

$$\left(\frac{(a_C)^r(a_D)^m}{(a_A)^p(a_B)^q}\right) = \text{Constant} = K_a \qquad ...(8)$$

where K_a is the equilibrium constant of the reaction. Equation (8) is the mathematical statement of the law of mass action.

Substituting equation (8) in (7), we get

$$\Delta G^o = -RT \text{ in } K_a \qquad ...(9)$$

This is another important relation.

VAN'T HOFF'S REACTION ISOTHERM

This can be deduced by either of the following methods:

1. Free Energy Change Method

In the earlier deduction it was assumed that the system remains in equilibrium and hence the change in free energy is zero. Now let us proceed to calculate the change in free energy when a chemical reaction is taking place out at constant temperature from some arbitrary concentrations of reactants to some other arbitrary concentrations of the products. Let us assume a reaction involving for gases, A, B, C and D

$$A + B \rightleftarrows C + D \qquad ...(1)$$

I this reaction amounts of A and B are decreasing while those of C and D are increasing. Subsequently, the free energies of A and B are decreasing while those C and D are increasing.

The free energy of the substance A per mole at temperature T is give as follows:

$$G_A = G_A^o + RT \ln p_A$$

where p_A represents the pressure of A and G_A^o represents the free energy at some standard state (p = 1) and is known as standard free energy.

Similarly, the free energies of B, C and D are as follows:

$$G_B = G_A^o + RT \ln p_B$$

$$G_C = G_C^o + RT \ln p_C$$

$$G_D = G_D^o + RT \ln p_D$$

By definition, the free energies of the reaction (ΔG) is

$$\Delta G = (\Delta G_C + G_D) + (G_A + G_B)$$

Hence

$$\Delta G = (G_C^o + RT \ln p_C + G_D^o + RT \ln p_D) - (G_A^o + RT \ln p_A + G_B^o + RT \ln p_B)$$

$$= (G_C^o + G_D^o - G_A^o - G_B^o) + RT \ln \frac{p_C \times p_D}{p_A \times p_B}$$

$$= \Delta G^o + RT \ln \frac{p_C \times p_D}{p_A \times p_B} \qquad ...(2)$$

where ΔG^o represents standard free energy of reaction.

But at equilibrium $\Delta G = 0$

Therefore, $\Delta G^o + RT \ln \left[\frac{p_C \times p_D}{p_A \times p_B}\right] = 0$

where pressure are equilibrium pressures,

or $\Delta G^o + RT \ln K_p = 0$

or $\Delta G^o = -RT \ln K_p$ where $K_p = \frac{p_C \times p_D}{p_A \times p_B}$

On substituting this value of ΔG^o in equation (2), we get

$$\Delta G = -RT \ln K_P + RT \ln \frac{p_C \times p_D}{p_A \times p_B}$$

or

$$-\Delta G = RT \ln K_P - RT \ln \frac{p_C \times p_D}{p_A \times p_B} \qquad ...(3)$$

Now consider the general reaction

$$n_1A + n_2B + ... \rightleftarrows^* n_3C + n_4D + ...$$

By following the same procedure, one may write

$$-\Delta G = RT \ln K_P + RT \ln \frac{(p_C)^{n_3} \times (p_D)^{n_4}}{(p_A)^{n_1} \times (p_B)^{n_2}} \qquad ...(4)$$

$$-\Delta G = RT \ln K_p - RT \Sigma \ln p \qquad ...(5)$$

Thus $$-\Delta G = RT \ln K_c - RT \Sigma \ln c \quad [\because \quad p \propto c] \quad ...(6)$$

where K_c is equilibrium constant. Equations (5) and (6) are known as Van't Hoff isotherm, because the reaction is taking place at constant temperature. This relation is very important because at may be used to show the direction in which a reaction tends to proceed.

2. The Van't Hoff's Equilibrium Box Method

It is a theoretical device which has been used successfully for calculating the free energy change in a chemical reaction taking place at constant temperature from arbitrarily chosen starting concentration of the reactants up to some of other arbitrarily chosen concentrations of the reactants. Consider the following chemical reaction.

$$n_1A + n_2B \rightarrow n_3C + n_4D \quad ...(1)$$

Let there be four reservoirs each of indefinite dimensions, one for each of the components A, B, C and D. Let the concentration, pressure and volume of A in the reservoir I be C_A, P_A and V_A, respectively. Similarly let C_B, P_B and V_B, C_C, P_C and VC and C_D, P_D and V_D be the corresponding concentrations, pressures and volumes of components B, C and D i the reservoirs II, I and IV respectively. Besides these vessels, there is a central reservoir which is an equilibrium box V having A, B C and D where their concentrations, partial pressures and molar volumes are $(C_A)_e$, $(P_A)_e$ and $(V_A)_e$, $(C_B)_e$, $(P_B)_e$ and $(V_B)_e$ $(C_C)_e$, $(P_C)_e$ $(P_C)_e$ and $(C_D)_e$, $(P_D)_e$ and $(V_D)_e$ respectively. Let us assume the temperature to be T. Now transfer n_1 moles of A from reservoir I into the equilibrium box V. This can be reversibly done by the following three stage process:

(1) The n_1 moles of A at pressure P_A and temperature T are taken out by slowly moving a piston of reservoir I. Thus, the necessary amount of A will come out through a wall which is permeable to A. It will involve work done by the gas, equivalent to n_1, P_A, V_A, *i.e.*

$$(\text{work done})_1 = n_1(P_AV_A) \quad ...(2)$$

(2) Before the gas has been introduced into the equilibrium reservoir V, the pressure of A is changed from P_A to $(P_A)_e$. The work done in the process for n_1 moles is

$$(\text{work done})_2 = n_1RT \ln \frac{P_A}{(P_A)_e} \quad ...(3)$$

(3) In the last stage, the gas is compressed at a pressure of $(P_A)_e$ into the equilibrium box. This involves work done which is equal to

$$(\text{work done})_3 = -n_1(P_A)_e(V_A)_e \quad \text{...(4)}$$

The total work done is, therefore, equal to sum of the equations (2), (3) and (4), *i.e.*,

$$w_1 = (\text{work done})_1 + (\text{work done})_2 + (\text{work done})_3$$

or
$$w_1 = n_1 P_A V_A + n_1 RT \ln \frac{P_A}{(P_A)_e} - n_1 (P_A)_e (V_A)_e \quad \text{...(5)}$$

But the gas is ideal and the process is isothermal. Therefore,

$$n_1 P_A V_A = n_1 (P_A)_e (V_A)_e \quad \text{...(6)}$$

On substituting equation (6) in (5), we obtain

$$w_1 = n_1 P_A V_A + n_1 RT \ln \frac{P_A}{(P_A)_e} - n_1 P_A V_A$$

$$= n_1 RT \ln \frac{P_A}{(p_A)_e} \quad \text{...(7)}$$

Similarly, if n_2 moles of B are transferred from reservoir II to the equilibrium box V the work done is given as follows:

$$w_2 = n_2 RT \ln \frac{p_B}{(p_B)_e} \quad \text{...(8)}$$

When n_1 moles of A and n_2 moles of B have bee introduced inside the equilibrium box they will react to form equilibrium concentrations of C and D. We have now to remove n_3 moles of C and n_4 moles of D from the equilibrium box V into the respective reservoirs III and IV. The work w_3 ad w_4 involved in these processes are as follows:

$$w_3 = n_3 RT \ln \frac{(p_C)_e}{P_C} \quad \text{...(9)}$$

$$w_4 = n_4 RT \ln \frac{(p_D)_e}{P_D} \quad \text{...(10)}$$

Total work done in this process is given as follows:

$$W = w_1 + w_2 + w_3 + w_4$$

$$W = n_1 RT \ln \frac{P_A}{(P_A)_e} + n_2 RT \ln \frac{P_B}{(P_B)_e}$$

$$+ n_3 RT \ln \frac{(P_C)_e}{P_C} + n_4 RT \ln \frac{(P_D)_e}{P_D} \quad \text{...(11)}$$

The work W is derived from the free energy decrease of the system *i.e.*

$$W = -\Delta G \qquad ...(12)$$

From equations (11) and (12), we have

$$-\Delta G = n_1RT \ln \frac{P_A}{(P_A)_e} + n_2RT \ln \frac{P_B}{(P_B)_e}$$

$$+ n_3RT \ln \frac{(P_C)_e}{P_C} + n_4RT \ln \frac{(P_D)_e}{P_D} \qquad ...(13)$$

$$-\Delta G = RT \ln \frac{(P_C)_e^{n_3}(P_D)_e^{n_4}}{(P_A)_e^{n_1}(P_B)_e^{n_2}} - RT \ln \frac{(P_C)^{n_3}(P_D)^{n_4}}{(P_A)^{n_1}(P_B)^{n_2}}$$

$$= RT \ln K_p - RT \Sigma n \ln P$$

Where $K_p = \dfrac{(P_C)_e^{n_3}(P_D)_e^{n_4}}{(P_A)_e^{n_1}(P_B)_e^{n_2}}$

and $\Sigma \ln n \; P = \ln(P_C)\, n_3 + \ln (P_D) n_4 - (\ln P_A)\, n_1 - (\ln P_B) n_2$

If, however, the work done is expressed in terms of C instead of P we would get

$$-\Delta G = RT \ln \frac{(C_C)_e^{n_3}(C_D)_e^{n_4}}{(C_A)_e^{n_1}(C_B)_e^{n_2}} - RT \ln \frac{(C_C)^{n_3}(C_D)^{n_4}}{(C_A)^{n_1}(C_B)^{n_2}}$$

$$= RT \ln K_c - RT \Sigma n \ln C \qquad ...(14)$$

where the term $\Sigma n \ln C$ has a similar significance as $\Sigma n \ln P$ already referred to.

VAN'T HOFF'S REACTION ISOCHORE

Van't Hoff's reaction isochore deals with the variation of equilibrium constant with temperature. According to Van't Hoff's reaction isotherm, the change in free energy is given as follows:

$$\Delta G = RT \ln K_c - RT \Sigma \ln C \qquad ...(1)$$

The Van't Hoff's isochore can be derived by differentiating equation (1) of Van't Hoff's reaction isotherm with respect to temperature at constant volume. But at constant volume the change in free energy is not given by ΔG but by ΔA. Therefore, equation (1) modifies to

$$-\Delta A = RT \ln K_c - RT \Sigma \ln c \qquad ...(2)$$

On differentiating it with respect to temperature at constant volume, we obtain

$$\left(\frac{\partial(-\Delta A)}{\partial T}\right)_V = RT\frac{d}{dT}(\ln K_c) + R\ln K_c$$

$$-RT\frac{d}{dT}(\Sigma n \ln c) + R\ln c$$

$RT\frac{d}{dT}\Sigma_n \ln c = 0$ because this arbitrary concentration is not considered as a function of temperature. Thus,

$$\left(\frac{\partial(-\Delta A)}{\partial T}\right)_V = RT\frac{d}{dT}(\ln K_c) + R\ln K_c - R\,\Sigma n \ln c$$

or $$-\frac{\partial(\Delta A)}{\partial T} = RT\frac{d}{dT}(\ln K_c) + R\ln K_c - R\,\Sigma n \ln c$$

or $$-\frac{\partial(\Delta A)}{\partial T} = RT\frac{d}{dT}(\ln K_c) + R\ln K_c - R\,\Sigma n \ln c$$

or $$-T\frac{\partial(\Delta A)}{\partial T} = RT^2\frac{d}{dT}(\ln K_c) + R\ln K_c - R\,\Sigma n \ln c$$

$$= RT^2\frac{d}{dT}(\ln K_c) - \Delta A \quad ...(3)$$

[use equation (2)]

The Gibb's Helmholtz equation is as follows:

$$\Delta A = \Delta E + T\left[\frac{\partial(AA)}{\partial T}\right]_V$$

or $$-T\left[\frac{\partial(AA)}{\partial T}\right]_V = \Delta E - \Delta A \quad ...(4)$$

On comparison of equation (3) with (2), we get

$$\Delta E = RT^2\frac{d}{dT}\ln K_c$$

or $$\frac{\Delta E}{RT^2} = \frac{d}{dT}\ln K_c \quad ...(5)$$

But $q_v = \Delta E$ where q_v is the heat absorbed by the system at constant volume. Hence we get

$$\frac{d}{dT}\ln K_c = \frac{q_v}{RT^2} \quad ...(6)$$

It is the required equation.

Integrated Form of the Van't Hoff Isochore

The integrated form of this can be obtained by integrating equation (5) between the temperatures T_1 and T_2 then

$$\int_1^2 d\ln K_c = \int_{T_1}^{T_2} \frac{\Delta E}{RT^2} dT$$

$$\frac{K_{c_2}}{K_{c_1}} = \frac{\Delta E}{R}\left[\frac{1}{T_1} - \frac{1}{T_2}\right] \quad ...(7)$$

This equation is useful for calculating the value of equilibrium constant at any temperature provided equilibrium constant at one temperature and ΔE are given. Even ΔE can also be calculated when K_{c_1} and K_{c_2} at two temperature T_1 and T_2 are known.

VAN'T HOFF REACTION ISOBAR

Some confusion exists in the literature about the reaction isobar. Some prefer to call it by the name of Van't Hoff reaction Isotherm. We will now deduce this equation by using the following relation.

$$K_P = K_c(RT)^{\Delta n}$$

where Δn represents change in number of moles during the reaction.

On taking logarithms of both sides, we get

$$\ln K_P = \ln K_c + \Delta n \ln RT$$

or

$$\ln K_P = \ln K_c + \Delta n \ln R + \Delta n \ln T$$

On differentiating this equation with respect to temperature we obtain

$$\frac{d}{dT}(\ln K_P) = \frac{d}{dT}(\ln K_c) + \frac{d}{dT}(\Delta n \ln R) + \frac{d}{dT}(\Delta n \ln T)$$

But $\frac{d}{dT}[\Delta n \ln R] = 0$, R is a constant

$$\therefore \quad \frac{d}{dT}(\ln K_P) = \frac{d}{dT}(\ln K_c) + \frac{\Delta n}{T} \qquad \left(\because \frac{d}{dT}\ln T = \frac{1}{T}\right)$$

But $\frac{d}{dT}(\ln K_c) = \frac{q_v}{RT^2}$ [From reaction isochore]

$$\therefore \quad \frac{d}{dT}(\ln K_c) = \frac{q_v}{RT^2} + \frac{\Delta n}{T} = \frac{q_v + \Delta n RT}{RT^2} \quad ...(1)$$

For an ideal as,

$$P\Delta V = \Delta n\ RT$$

Therefore, equation (1) becomes as follows:

$$\frac{d}{dT}(\ln K_P) = \frac{q_v + P\Delta V}{RT^2}$$

But $q_p = q_v + P\Delta V$, where q_p represents the heat absorbed by the system at constant pressure. Hence,

$$\frac{d}{dT}(\ln K_P) = \frac{q_p}{RT^2} \quad ...(2)$$

We also know

$$\Delta H = q_p$$

$$\therefore \quad \frac{d}{dT}(\ln K_P) = \frac{\Delta H}{RT^2} \quad ...(3)$$

This equation is known as Van't Hoff's reaction isobar. By integrating equation (3), we obtain

$$\int_1^2 d \ln K_P = \int_{T_1}^{T_2} \frac{\Delta H}{RT^2} dH$$

$$\frac{K_{P_2}}{K_{P_1}} = \frac{\Delta H}{R}\left[\frac{1}{T_1} - \frac{1}{T_2}\right]$$

Knowing the equilibrium constant K_{P_2} at temperature T_2 it is possible to calculate the equilibrium constant K_{P_1} at temperature T_1 and *vice versa*, provided the heat of reaction ΔH is known. Alternatively if the equilibrium constants of a reaction at two temperatures are known, the heat of reaction (ΔH) can be calculated.

APPLICATIONS OF LAW OF MASS ACTION

1. Dissociation of Hydrogen iodide

This is first type (*i.e.*, $K_p = K_c$) of a homogeneous gaseous reversible reaction as in this case the total number of molecules on both sides is the same.

$$2HI \rightleftarrows^{*} H_2 + I_2 - \text{Heat}$$

	2HI	H_2	I_2
Initial	1 mole	0	0
At equilibrium	(1 – x) mole	x/2 mole	x/2 mole

Concentration $\frac{(1-x)}{V}$ $\frac{x}{2V}$ $\frac{x}{2V}$

(where V = volume of vessel)

Let us start with '1' g mole of HI enclosed in a vessel of volume V litres. Let 'x' mole of HI decompose at the equilibrium to yield x/2 mole of each H_2 and I_2. Then the equilibrium molar concentrations are:

$$[HI] = \frac{(1-x)}{V};$$

$$[H_2] = \frac{x/2}{V}; \quad [I_2] = \frac{x/2}{V}$$

$\therefore$ From the Law of Mass action, we get

$$K_P = K_C = \frac{k_f}{k_b} = \frac{[H_2][I_2]}{[HI]} = \frac{\frac{x}{2V}\cdot\frac{x}{2V}}{\left[\frac{1-x}{V}\right]^2} = \frac{x^2}{4(1-x^2)}$$

(1) *Effect of pressure change,* From the above equation it is clear that K_c is *independent* of total volume 'V' of the system. A change of pressure alters the total volume and as K_c is independent of 'V' consequently *change of pressure will not alter the final state of equilibrium.*

(2) *Effect of temperature,* With the rise and fall of temperature the velocity constants k_f and k_b vary differently. It has been found that k_f increases more rapidly than k_b with the increase of temperature. So, with the increase of temperature $K_c = k_f/k_b$ will increase. Evidently, the concentration of products (*i.e.* H_2 and I_2) will increase and that of reactant HI will decrease. *Hence, the temperature will favour the dissociation of HI.*

(3) *Effect of adding products,* Alternative equation for equilibrium constant K_P in terms of partial pressures, will be:

$$K_P = \frac{P_{H_2} \times P_{I_2}}{P_{HI}^2}$$

On adding of any product, say H_2 the value of P_{H_2} will increase. In order to keep the value of K_P constant, the value of P_{I_2} must decrease and of P_{HI} must increase. In other words, *adding of H_2 will suppress the dissociation of HI.* Similar will be the consequences if I_2 is added.

(4) *Effect of adding an inert gas:* Addition of an inert gas like argon to the equilibrium mixture will increase the total pressure of the reactants without change the partial pressures of H_2, I_2 or HI. Hence, *the addition of an inert gas will have no effect on the degree of dissociation of HI.*

2. Dissociation of PCl_5

It is a second type (*i.e.*, $K_p \neq K_c$) of a homogeneous gaseous reaction as the total number of molecules of the reactants *change* as a result of chemical reaction.

	$PCl_5 \rightleftarrows^*$	PCl_3 +	Cl_2
Initial	1 mole	0	0
At equilibrium	(1 – x) mole	x mole	x mole
			= 1 – x + x + x = 1 + x mole
Concentration	$\frac{(1-x)}{V}$	$\frac{x}{V}$	$\frac{x}{V}$
			where V = Volume
Partial pressure	$\frac{(1-x)}{(1+x)}P$	$\frac{xP}{(1+x)}$	$\frac{xP}{(1+x)}$
			where P = Pressure

Let us start with '1' g mole of PCl_5 enclosed in a vessel of volume 'V' litres. Let 'x' mole of PCl_5 decompose at the equilibrium to yield x mole of each PCl_3 and Cl_2. Then, equilibrium molar concentrations are:

$$[PCl_5] = \frac{1-x}{V}$$

$$[PCl_3] = [Cl_2] = \frac{x}{V}$$

∴ Applying Law of Mass Action we get

$$\{K_c = \frac{k_f}{k_b} = \frac{[PCl_3][Cl_2]}{[PCl_5] = \frac{x^2}{(1-x)V}} \quad ...(i)$$

The expression for K_p can be deduced as follows:

Total No. of moles at equilibrium = 1 – x + x + x = (1 + x). Let P = Total pressure of the system,

Then $$P_{PCl_5} = \frac{1-x}{1+x}P; \quad P_{PCl_3} = \frac{xP}{1+x};$$

$$P_{PCl_2} = \frac{xP}{1+x}$$

$$\therefore \quad K_p = \frac{p_{PCl_3} \times p_{Cl_2}}{p_{PCl_5}}$$

$$= \frac{x^2P}{(1-x)(1+x)} = \frac{x^2P}{(1-x^2)} \quad ...(ii)$$

(1) *Effect of pressure change:* In the expression (ii) for K_p, p is in the numerator. Hence, it follows if P is increased, the value of 'x' must decrease i order to keep K_p constant. Evidently increase in pressure suppresses the dissociation of PCl_5.

(2) *Effect of temperature:* It has been found that k_f *increases more rapidly than* k_b *with the rise of temperature.* Thus, it is evident that with the increase of temperature the value of $K = k_f/k_b$ increases, *i.e.*, the concentration of products increases and that of reactant PCl_5 decreases. *Hence, high temperature favours the dissociation of* PCl_5.

(3) *Effect of adding products at constant volume:* We have:

$$K_p = \frac{p_{PCl_3} \times p_{Cl_2}}{p_{PCl_5}}$$

Addition of any product, say Cl_2 will result in the increase in the value of PCl_3. In order to keep K_p constants, the value of PCl_3 will decrease and that of PCl_5 would decrease. Evidently, *addition of* Cl_2 *suppresses the dissociation of* PCl_5. Similarly consequences will happen by addin PCl_3.

(4) *Effect of adding an inert gas at constant volume:* The addition of an inert gas will increase the total pressure P, since K_p is proportional to P. Hence, in order to keep K_p constant the increase of P by the addition of inert gas will decrease the value of x. Evidently, *addition of an inert as will support the dissociation of* PCl_5.

3. Dissociation of N_2O_4

This is a second type (*i.e.*, $K_P \neq K_C$) of homogeneous gaseous reaction as the number of molecules alters as a result of reaction.

	$N_2O_4 \rightleftarrows$*	$2NO_2$
Initial	1 mole	0

At equilibrium(1 – x) mole 2x mole

Total = 1 – x + 2 = 1 + x moles

Concentration $\frac{(1-x)}{V}$ $\frac{2x}{V}$ where V = Volume

Partial pressure $\frac{(1-x)}{(1+x)}P$ $\frac{2xP}{(1+x)}$ where P = Pressure

Let us start with 1 mole of N_2O_4 enclosed in a vessel of capacity 'V' litres. Let at equilibrium 'x' mole of N_2O_4 is dissociated into 2x moles of NO_2. Then, equilibrium concentrations are:

$$[N_2O_4] = \frac{1-x}{V};$$

$$[NO_2] = \frac{2x}{V}$$

∴ Applying Law of Mass Action, we get

$$K_c = \frac{k_f}{k_b} = \frac{[NO_2]^2}{[N_2O_4]} = \frac{4x^2}{(1-x)V} \quad \text{...(i)}$$

The expression for K_P can be calculated as follows:

Total No. of moles at equilibrium = 1 – x + 2x = 1 + x. Let total pressure of the system = P.

Then, $$p_{N_2O_4} = \frac{(1-x)}{(1+x)}P;$$

$$p_{NO_2} = \frac{2xp}{1+x}$$

$$\therefore \quad K_P = \frac{(p_{NO_2})^2}{p_{N_2O_4}} = \frac{4x^2P}{(1-x^2)} \quad \text{...(ii)}$$

(1) *Effect of pressure change:* From the expression (ii) it is clear that when P increases x must decrease and *virsa* in order to keep K_P constant. In other words, *increase of pressure suppresses the dissociation of N_2O_4.*

(2) *Effect of adding product at constant volume:* We have

$$K_P = p^2_{NO_2} / p_{N_2O_4}$$

If NO_2 is added to the equilibrium mixture in a close vessel, the partial pressure p_{NO} will increase. In order to keep K_P constant, the value of $P_{N_2O_4}$ must increase. Hence, *dissociation of N_2O_4 will be suppressed by the addition of NO_2.*

(3) *Effect of change of temperature:* It has been found that with the *rise of temperature k_f increases more rapidly than k_b, i.e., K increases* or the concentration of product NO_2 will increase.

Hence, *increase in temperature will favour the formation of NO_2, i.e., dissociation of N_2O_4.*

(4) *Effect of adding an inert gas at constant volume:* The addition of an inert gas will increase the total pressure P, since K_P, is directly proportional to P.

Hence, in order to keep K_P constant, the increase of total pressure P by the addition of inert gas will decrease the value of 'x'. Evidently, *the addition of an inert gas at constant volume will suppress the dissociation of N_2O_4.*

4. Synthesis of Ammonia

This is a second type (*i.e.*, $K_P \neq K_C$) of homogeneous gaseous reaction as the total number of molecules *changes* as a result of reaction

$$N_2 + 3H_2 \rightleftarrows^{*} 2NH_3$$

	N_2	H_2	NH_3
Initial	1 mole	3 moles	0
At equilibrium	(1 – x) mole	(3 – 3x) moles	2x mole

= 1 – x + 3 – 3x + 2x = 4 – 2x

Concentration	$\frac{(1-x)}{V}$	$\frac{(3-3x)}{V}$	$\frac{2x}{V}$

where V = Volume

Partial pressure	$\frac{(1-x)P}{(4-2x)}$	$\frac{(3-3x)P}{(4-2x)}$	$\frac{2xP}{(4-2x)}$

where P = Pressure

Let us start with 1 mole of N_2 and 3 moles of H_2 in a closed vessel of volume 'V' litres. Let 'x' mole of N_2 and 3x moles of H_2 are used up at equilibrium. Then, equilibrium concentrations are:

$$[N_2] = \frac{1-x}{V}; \quad [H_2] = \frac{3(1-x)}{V}; \quad [NH_3] = \frac{2x}{V}$$

∴ Applying law of Mass Action, we get

$$K_C = \frac{k_f}{k_b} = \frac{[NH_3]^2}{[N_2][H_2]^3} = \frac{\left[\frac{2x}{V}\right]^2}{\left[\frac{1-x}{V}\right]\left[\frac{3(1-x)}{V}\right]^3} = \frac{4x^2V}{27(1-x)^4} \quad ...(i)$$

The expression for K_p can be calculated as follows:

Total No. of molecules in the system at equilibrium

$$= 1 - x + 3 - 3x + 2x = 4 - 2x.$$

Let total pressure of system = P.

Then $p_{N_2} = \frac{1-x}{4-2x}P;$

$$p_{H_2} = \frac{3-3x}{4-2x}P;$$

$$p_{NH_3} = \frac{2xP}{4-2x}$$

$$\therefore \quad K_p = \frac{(p_{NH_3})^2}{p_{N_2} \times (p_{H_2})^3}$$

$$= \frac{\left[\frac{2xP}{4-2x}\right]^2}{\left[\frac{1-x}{4-2x}P\right]\left[\frac{(3-3x)P}{4-2x}\right]^3} = \frac{16x^2(2-x)^2}{27P^2(1-x)^4} \quad ...(ii)$$

1. *Effect of pressure change:* As K_p is inversely proportional to P^2, so an increase in P will cause a decrease in x. In other words, *increase of pressure increases the formation of NH_3.*

2. *Effect of temperature change:* It has bee found that *with the rise of temperature k_b increases more rapidly than k_f.* Thus, with the increase in temperature the backward reaction is favoured. Hence, at high temperature the dissociation of NH_3 will occur. Conversely *low temperature favours the formation of NH_3.*

3. *Effect of adding reactants at constant volume:* We have:

$$K_p = \frac{(p_{NH_3})^2}{p_{N_2} \times (p_{H_2})^3} m$$

Addition of any reactant, say N_2, will result ion the increase in the value of p_{N_2}. In order to keep K_P constant, the value of p_{NH_3} will increase and that of p_{H_2} will decrease. Evidently, addition of *N_2 favours the formation of NH_3*. Similar consequences will happen by adding H_2.

4. *Effect of adding an inert gas at constant volume:* The addition of an inert gas will decrease the total pressure P, since K_P is inversely proportional to P.

Hence, in order to keep K_P constant the increase of P by addition of inert gas will increase the value of x. Evidently, *addition of an inert gas will favour the formation of NH_3*.

5. Formation of Nitric Oxide

This is a first type (*i.e.*, $K_P = K_C$) of a homogeneous reaction as the total number of molecules remains the same as a result of the reaction:

	N_2 +	O_2 ⇄*	2NHO – Heat
Initial	a mole	b moles	0
At equilibrium	(a – x) mole	(b – x) moles	2x mole

= 1 – x + 3 – 3x + 2x = 4 – 2x

Concentration	$\frac{(a-x)}{V}$	$\frac{b-x}{V}$	$\frac{2x}{V}$

(where V = Volume of vessel)

Let us start with 'a' moles of N_2 and 'b' moles of O_2 in a closed vessel of capacity 'V' litres. Let at equilibrium 'x' moles each of N_2 and O_2 are used up to form O_2 moles of NO. Then, the equilibrium concentrations are:

$$[N_2] = (a - x)/V ;$$

$$[O_2] = (b - x)/V;$$

$$[NO] = 2x/V$$

∴ Applying Law of Mass Action, we get

$$K_P = K_C = \frac{[NO]^2}{[N_2][O_2]} = \frac{4x^2}{(a-x)(b-x)}$$

2. *Effect of adding reactants:* We have

$$K_P = \frac{p_{NO}^2}{p_{N_2} \times p_{O_2}}$$

of adding a reactant, say N_2 the partial pressure of N_2 will increase. In order to keep K_P constant, the value of p_{O_2} will decrease and that of p_{NO} will increase, *i.e.*, more N_2 and O_2 will react to form NO.

In other words, addition of N_2 will favour the formation of NO. Similar effects will be caused by the addition of oxygen.

3. *Effect of temperature change:* It has been found that with the rise of *temperature k_f increases more rapidly than k_b* *i.e.* with the increase of temperature the about of product, *i.e.*, NO, will increase.

Hence *formation of nitric oxide is favoured by high temperature.*

6. Esterification of Ethyl Alcohol by Acetic Acid

This is a reaction in solution and is represented by the equation.

	CH_3COOH +	C_2H_5OH ⇄	$CH_3COOC_2H_5$ +	H_2O
Initial	a mole	b moles	0	0
At equilibrium	(a – x) mole	(b – x) moles	x mole	x mole
Concentration	$\frac{(a-x)}{V}$	$\frac{b-x}{V}$	$\frac{x}{V}$	$\frac{x}{V}$

Let us start with 'a' moles of acid and 'b' moles of alcohol. Let 'x' moles each of ester and water are formed at equilibrium. If V is thee total volume of the solution mixture, then equilibrium molar concentrations are:

$$[CH_3COOH] = (a - x)/V;$$

$$[C_2H_5OH] = \frac{b-x}{V}; \; [CH_3COOC_2H_5] = x/V \; [H_2O] = x/V$$

∴ Applying Law of Mass Action, we get

$$K_C = \frac{k_f}{k_b}$$

$$= \frac{[CH_3COOC_2H_5][H_2O]}{[CH_3COOH][C_2H_5OH]}$$

$$= \frac{x/V \times x/V}{\frac{a-x}{V} \times \frac{b-x}{V}} = \frac{x^2}{(a-x)(b-x)}$$

LAW OF CHEMICAL EQUILIBRIUM AND HETEROGENEOUS EQUILIBRIA

When chemical equilibrium is established in one phase—a mixture of gases, a liquid solution - we have a case of *homogeneous equilibrium.* When more than one phases are involved—for example, gas and solid, or liquid and solid or liquid and gas equilibrium is said to be *heterogeneous.*

If a chemical equilibrium consists of two or more than two phases one of which at least must be a solid or liquid phase. It is said to be heterogeneous equilibrium.

For example, water in contact with its vapour in a closed space forms a two phase (Liquid-Gas) heterogeneous equilibrium.

A saturated solution in equilibrium with solute is another familiar instance of *heterogeneous equilibrium.*

Dissociation of red lead can similarly be regarded as a heterogeneous equilibrium consisting of three phases, two of which (Pb_3O_4 ad PbO) are solid and the third (O_2) a gas.

$$2Pb_3O_4(s) \rightleftarrows 6PbO(s) + O_2(g)$$

For the present we shall concern ourselves with gas-solid reactions which are the ore common.

(i) Thermal Decomposition of $CaCO_3$

A well known example of such a reaction is the dissociation of calcium carbonate:

$$CaCO_3(s) \rightleftarrows CaO(s) + CO_2(g)$$

where there is one gas and two solids. We could write the equilibrium constant as,

$$K_C = \frac{[CaO][CO_2]}{[CaCO_3]}$$

We now have to decide what is meant by the "concentrations" of the calcium carbonate and calcium oxide. These two solids will form two separate phases, since it is unusual for solids to mix. Any pure solid at a particular temperature and pressure, will contain a constant number of molecules per unit volume, so that the molar concentration of a solid is a constant.

Looking at it another way, we have seen that the vapour pressure of a liquid in a mixture is proportional to its molar concentration; the vapour pressure of pure liquid is constant, at a definite temperature, so that the concentration must be a constant. A solid also has a vapour pressure although it may be too small to be measured.

This vapour pressure is constant provided temperature is constant, irrespective of the amount of solid that may be present. By analog, we can use the vapour pressure of a solid as a measure of the concentration and we shall then arrive at the same result, *i.e.*, the pure solid has a constant concentration. The actual value that is given to these constants does not matter provided that we always use the same values. By convention, the *concentration of a pure liquid or a pure solid is taken as unity.* The actual amount present has no effect on the concentration.

The equilibrium constant for the calcium carbonate reaction now becomes.

$$K_C = [CO_2]$$

and since we are now dealing with the concentration of a gas, it is more convenient to use the pressure:

$$K_P = p_{CO_2}$$

The pressure of the CO_2 is also the total pressure of the system, since it is the only gas present. This is referred to as the *dissociation pressure* and since it is equal to K_P it follows that the dissociation pressure has definite value for any particular temperature.

The pressure of the system is fixed and cannot be altered without upsetting the equilibrium. Experimentally it has been found that the dissociation pressure of CO_2 in equilibrium with solid $CaCO_3$ and solid CaO, at any temperature, is constant.

Similar to homogeneous equilibria, the law of chemical equilibrium is also applicable to heterogeneous equilibria. The application of this law to the heterogeneous equilibria is based of the assumption that the reaction goes on i a heterogeneous system. Consider the above mentioned thermal dissociation of calcium carbonate.

$$CaCO_3(s) \rightleftarrows CaO(s) + CO_2(g)$$

We can assume that th reaction here is taking place between calcium carbonate vapour, calcium oxide vapour and carbon dioxide gas, the whole system is thus functioning like a homogenous system.

This assumption the leads us to further conclusion, that the solid $CaCO_3$ and the solid CaO, like other constituents also possess a definite vapour pressure and hence the equilibrium concentrations of the reacting substances can be expressed in terms of their partial pressures as in the case of other homogeneous equilibria. Therefore, p_{CaCO_3}, p_{CaO} and p_{CO_2}, respectively represent the partial pressures of $CaCO_3$, CaO and CO_2, then application of the law of chemical equilibrium gives the expression:

$$K_P = \frac{p_{CaO} \times p_{CO_2}}{p_{CaCO_3}}$$

The partial pressure of a solid (or a pure liquid) is very small but is constant at a particular temperature and remains so as ion as there is any solid (or liquid) present in the system p_{CaCO_3}, and p_{CaO} are thus constant. The equilibrium law equation, thus reduces to,

$$p_{CO_2} = K'_P$$

Thus, when calcium carbonate is heated in a closed vessel at a definite temperature, the pressure of CO_2 formed remains constant irrespective of the amount of calcium carbonate taken. Therefore at a given temperature, the pressure of CO_2 will be constant. The equilibrium constant K'_P, therefore, appears to be determined only by th pressure of CO_2. If CO_2 is allowed to escape, its pressure will decrease and further dissociation of calcium carbonate will take place until the original pressure is regarded. That is why in the line kilns, where the CO_2 escapes, the reaction proceeds practically in one direction, *i.e.*, the dissociation of $CaCO_3$,

The equilibrium pressure of CO_2 which is constant at a particular temperature is known as the dissociation pressure of $CaCO_3$. The dissociation pressure at a particular temperature will to change, even if any product of the reaction is added or removed from the system. It changes with the change of temperature.

(ii) Thermal Dissociation of Solid Ammonium Hydrrosulphide

The solid ammonium hydrosulphide, on heating dissociates into two gases, viz., ammonia and hydrogen sulphide $NH_4HS(s) \rightleftarrows NH_3(g) + H_2S(g)$. It is another example of heterogeneous equilibrium. Since the partial pressure of solid component, *i.e.*, NH_4HS will remain constant at a particular temperature, as ion as there is any solid left in the system th equilibrium constant KP can be expressed as:

$$K_p = \frac{p_{NH_3} \times p_{H_2S}}{p_{NH_4HS}} = \frac{p_{NH_3} \times P_{H_2S}}{\text{constant}}$$

where p_{NH_3} and p_{H_2S} represent the partial pressures of ammonia and hydrogen sulphide. Evidently

$$K'_P = p_{NH_3} \times p_{H_2S}$$

i.e. At any temperature the product of the partial pressures of ammonia and hydrogen sulphide (p_{NH_4HS}) being very small, is neglected), and hence each will be equal to p/2.

i.e.

$$p_{NH_3} \times p_{H_2S} = p/2$$

$$K'_P = p_{NH_3} \times p_{H_2S} = p/2 \times p/2 = p^2/4$$

Thus, from the total pressure of the system which is constant at a given temperature, the value K'_P can be determined. If one of the two gases is introduced in the system, from outside at equilibrium, p_{NH_3} will no longer be equal to p_{H_2S}, but the product $p_{NH_3} \times p_{H_2S}$ will still remain constant. Consequently, the dissociation of the solid will be diminished and a certain amount of solid hydrosulphide will be deposited. Isambart confirmed the above observations by investigating the dissociation equilibrium of ammonium hydrosulphide and ammonium cyanide.

(iii) Iron-Steam Reaction

Another type of heterogeneous equilibrium is the reaction of steam with iron:

$$3Fe(s) + 4H_2O(g) \rightleftarrows Fe_3O_4(s) + 4H_2(g)$$

The equilibrium constant is given by

$$K_c = \frac{[Fe_3O_4][H_2]^4}{[Fe]^3 \times [H_2O]^4}$$

If the active asses of solids are taken constant, then

$$K' = \frac{[H_2]^4}{[H_2O]^4}$$

or

$$K'' = \frac{[H_2]}{[H_2O]}$$

or $$K_P = \frac{p_{H_2}}{p_{H_2O}}$$

Experimentally also it has been found that the ratio p_{H_2}/p_{H_2O} is constant at any one temperature and independent of amounts or nature of iron and iron oxide present.

(iv) Water Gas Reaction

In this reaction steam is passed over heated coke:

$$H_2O(g) + C(s) \rightleftarrows CO(g) + H_2(g)$$

for which the equilibrium constant is

$$K_P = \frac{p_{CO} \times p_{H_2}}{p_{H_2O}}$$

as the partial pressure of carbon is taken as unity.

SOLVED EXAMPLES

Example 1:

Show that the equation, v = u + at is dimensionally correct and the equation v = u + at² dimensionally incorrect. Here, u is the initial velocity, v is the final velocity, a is Laws of Chemical Combination and t is the time interval.

Solution:

We know (from the previous section),

$$[v] = [u] = L/T$$

and $$[a] = L/T^2$$

Then $$[at] = L/T^2 \times T = L/T$$

Thus, the dimensions of the quantities on both the sides of the equation, v = u + at, are the same. Therefore, the given equation is dimensionally correct.

For the equation, $v = u + at^2$

$$[v] = \frac{L}{T} \quad [u] = \frac{L}{T}$$

and $$[a] = \frac{L}{T^2}$$

Then $$[at^2] = \frac{L}{T^2} \times T^2 = L$$

In this case, the dimensions of the quantities on the two sides of the dimensional equation are not the same. Therefore, the equation, v = u + at^2 is dimensionally incorrect.

Example 2:

Convert 1.5 minutes into seconds.

Solution:

We know that 1 min = 60 s

So, $1 = \frac{60\,s}{1\,min}$ (note that the left hand side is a dimensionless quantity)

We can now write,

$$1.5 \text{ min} = 1 \times 1.5 \text{ min} = \frac{60\,s}{1\,min} \times 1.5 \text{ min} = 1.5 \times 60 \text{ s} = 90 \text{ s}$$

Example 3:

How many milligrams of bromine are there in a bottle containing 1/10 kg of this substance?

Solution:

We know that, 1 kg = 1000 g

and $1 \text{ mg} = 10^{-3} \text{ g}$

or $1 \text{ g} = 10^3 \text{ g}$

These relationships give,

$$\frac{1}{10}\text{kg} = \frac{1}{10} \times (1\,\text{kg}) = \frac{1}{10} \times (1000\,\text{g})$$

$$= \frac{1000}{10} \times (1\,\text{g}) = 100 \times \left(10^3\,\text{mg}\right) = 10^5 \text{ mg}$$

Thus, 1/10 kg equals 10 mg (or 1 lakh milligram).

Example 4:

Conventional unit of pressure is atm. Convert 1 atm pressure to its value in SI units.

Solution:

We know,

Pressure = Height × Density × Laws of Chemical Combination due to gravity = h × d × g

In CGS system of units,

$$1 \text{ atm} = (76 \text{ cm}) \times (13.6 \text{ g/cm}^3) \times (980.67 \text{ cm/s}^2)$$

$$= (76 \times 13.6 \times 980.67 \left(\text{cm} \times \frac{\text{g}}{\text{cm}^3} \times \frac{\text{cm}}{\text{s}^2} \right)$$

$$= 1.0136 \times 10^6 \left(\frac{1\text{g}}{1\text{cm} \times 1\text{s}^2} \right) \quad \text{...(1)}$$

Since , g and cm are not SI units, hence these should be converted to the appropriate SI unit.

Therefore,

$$\frac{1\text{g}}{1\text{cm} \times 1\text{ s}^2} = \frac{10^{-3}\text{ kg}}{10^{-2}\text{ m} \times 1\text{s}^2} = \frac{1}{10}\left(\frac{1\text{kg}}{1\text{m} \times 1\text{s}^2} \right)$$

$$\begin{bmatrix} 1\text{ g} = 10^{-3}\text{ kg} \\ 1\text{cm} = 10^{-2}\text{ m} \end{bmatrix} \quad \text{...(2)}$$

$$1 \text{ atm} = 1.0136 \times 10^6 \times \frac{1}{10} \left(\frac{1\text{kg}}{1\text{m} \times 1\text{s}^2} \right) = 1.0136 \times 105\text{kg m}^{-1}\text{ s}^2$$

$$= 1.0136 \times 10^5 \times (1 \text{ kg m}^{-1} \text{ s}^{-2})$$

But, $1 \text{ kg m s}^{-2} = 1 \text{ N}$

Then, $1 \text{ atm} = 1.0136 \times 10^5 \text{ N m}^{-2} = 1.0136 \times 10^5 \text{ Pa}$

Therefore,

$$1 \text{ atm} = 1.0136 \times 10^5 \text{ Pa} = 1.0136 \times 10^5 \text{ kg m}^{-1} \text{ s}^{-2}$$

Example 5:

Express each of the following in the SI units.

(i) 93 million miles (this is the distance between earth and sun)

(ii) 5 feet and 2 inches (the average height of an Indian female)

(iii) 100 miles per hour (the typical speed of Rajdhani Express)

(iv) 14 pounds per sq inch (atmospheric pressure)

(v) 0.74 Å (the bond length of hydrogen molecule)

(vi) 46°C (the peak summer temperature in Delhi)

(vii) 150 pounds (the average weight of an Indian male).

Solution:

(i) 93 million miles = 93×10^6 miles

But, 1 mile = 1.60934 km = 1.60934×10^3 m

So, $$1 = \frac{1.60934 \times 10^3 \text{ m}}{1 \text{ mile}}$$

Then, 93 million mile = 93×10^6 mile = 93×10^6 mile × 1

$$= 93 \times 10^6 \text{ mile} \times \frac{1.60934 \times 10^3 \text{ m}}{1 \text{ mile}} = 1.5 \times 10^{11} \text{ m}$$

(ii) 5 feet and 2 inch = (5 × 12 + 2) inch = 62 inch

But, 1 inch = 2.54×10^{-2} m

$$1 = \frac{2.54 \times 10^{-2} \text{ m}}{1 \text{ inch}}$$

Then, 5 feet and 2 inch = 62 inch = 62 inch × 1

$$= 62 \text{ inch} \times \frac{2.54 \times 10^{-2} \text{ m}}{1 \text{ inch}} = 1.58 \text{ m}$$

(iii) 100 miles per hour = 100 ×

$$= 100 \times \frac{1.60934 \times 10^3 \text{ m}}{(60 \times 60\text{s})} = 44.7 \text{ m s}^{-1}$$

$$\left[\begin{array}{l} \because 1 \text{ mile} = 1.60934 \times 10^3 \text{ m} \\ 1 \text{ min} = 60 \times 60 \text{ s} \end{array}\right]$$

(iv) 14 pound per sq inch $= \dfrac{14 \times (1 \text{ lb force})}{(1 \text{ inch})^2}$

But, 1 lb force = 4.448 N

and 1 inch = 2.54×10^{-2} m

14 lb $(\text{inch})^2$

$$= \frac{14\,lb}{(1\,inch)^2} = \frac{14\,lb}{1\,inch^2}$$

$$= \frac{14 \times (4.448\,N)}{\left(2.54 \times 10^{-2}\,m\right)^2} = \frac{62.272\,N}{6.45 \times 10^{-4}\,m^2}$$

(v) 0.74Å

We know, $1Å = 10^{-10}$ m

or $1 = 10^{-10}$ m/1Å

Therefore, $0.74Å = 0.74Å \times 1 = \frac{0.74Å \times 10^{-10}\,m}{1\,Å}$

$= 0.74 \times 10^{-10}\,m = 7.4 \times 10^{-11}\,m$

(vi) $46°C = (46 + 273.15)\,k = 319.15\,K$

(vii) 150 pounds (note, it is the mass pound)

We know, 1 pound = 0.4536 kg

Therefore, 150 pound = 150 × (1 lb) = 150 × (0.4536 kg)

= 68.04 kg = 68.0 kg

Example 6:

Light radiation travel with a speed 3×10^8 m s^{-1}. Express the speed in miles per hour.

Solution:

Given: speed of light $= 3 \times 10^8\,m\,s^{-1} = 3 \times 10^8 \times \frac{(1m)}{(1s)}$

From the conversion factor table,

$1\,m = 6.2 \times 10^{-14}$ mile and $1\,s = \frac{1}{3600}\,h$

So, Speed of light $= 3 \times 10^8 \times \frac{6.2 \times 10^{-4}\,mile}{(1/3600)h}$

$= 3 \times 10^8 \times 6.2 \times 10^{-4} \times 3600$ mile/h

$= 6.7 \times 10^8$ mile/h.

Example 7:

Add the numbers, 11.1, 2.11 and 0.111 and report the final result to the correct place of decimal.

Solution:

The number of decimal places and the number of significant figures for each of the given number, and their sum are given below.

Given numbers	*No. of decimal places*	*No. of significant figures*	*Minimum no. of decimal places*	*Correctly reported final result*
11.1	1	3	-	
+ 2.11	2	3	-	
+0.111	3	3	-	
Sum = 13.32)	–	–	1	13.3

Therefore, the sum of the given numbers reported to the correct places of decimal is 13.3.

Example 8:

Add the numbers, 28.521, 6.38, 0.216 and 111.535 and report the final sum to the correct place of decimal.

Solution:

The number of decimal places and the number of significant figures for each of the given number and their sum are given below.

Given numbers	*No. of decimal places*	*No. of significant figures*	*Minimum no. of decimal places*	*Correctly reported final result*
28.521	3	5		
6.38	2	3		
0.216	3	3		
111.535	3	6		
Sum = 146.652				2146.65

The correct reporting of the sum is 146.65.

Example 9:

Subtract 4.2563 from 16.182 and report the result correctly.

Solution:

The given numbers, number of decimal pláces, and significant figures for each of the given number, and the difference between the two are given below.

Given numbers	*No. of decimal places*	*No. of significant figures places*	*Minimum no. of decimal*	*Correctly reported final result*
16.182	3	5		
–4.2563	4	5		
11.9257			3	11.926

Example 10:

Subtract 0.0017 from 9.64, and report the result correctly.

Solution:

The given numbers, number of decimals, and significant figures for each of the given number, and the difference between the two are given below.

Given numbers	*No. of decimal places*	*No. of significant figures*	*Minimum no. of decimal places*	*Correctly reported final result*
9.64	2	3		
–0.00017	4	2		
Difference:	9.6383		2	9.64

↑

the final result has five significant figure (Example 7):

Add the numbers, 11.1, 2.11 and 0.111 and report the final result to the correct place of decimal.

Solution:

The number of decimal places and the number of significant figures for each of the given number, and their sum are given below.

Given numbers places	*No. of decimal figures*	*No. of significant decimal places*	*Minimum no. of final result*	*Correctly reported*
11.1	1	3	-	
+ 2.11	2	3	-	
+0.111	3	3	-	
Sum = 13.32)	–	–	1	13.3

Therefore, the sum of the given numbers reported to the correct places of decimal is 13.3.

Example 11:

Multiply the numbers 0.256, 0.08205 and 298.16, and report the result correctly.

Solution:

Multiplication of the three numbers gives,

$$0.256 \times 0.08205 \times 298.16 = 6.26279$$

The number of significant figures in each of the terms involved in multiplication is

Quantity	*No. of significant figures*	
0.256	3	←this term has the least number of significant figures
0.08205	4	
298.16	5	

Since, here the least number of significant figures is 3, hence the final result should be reported up to only 3 significant figures.

This is done by rounding off the term 6.26279 by following the rules of rounding off. The rounding off the term 6.26279 to three significant figures gives 6.26.

Example 12:

Divide 0.36 by 2.487, and report the result correctly.

Solution:

Division of 0.36 by 2.487 gives, 0.36 + 2.487 = 0.14475

Since, the least number of significant figures in 0.36 is 2, hence the final result will be reported only up to two significant figures.

This is done by rounding off the term 0.14475. Rounding off this number to two significant figures gives 0.14. Thus, the correctly reported result is 0.14.

Example 13:

Solve the expression,

5.28 × 0.156 × 3 + 0.0428 and report the correct result.

Solution:

The solution obtained on calculator is,

$$5.28 \times 0.156 \times 3 + 0.0428 = 57.734579$$

The number of the significant figures in various terms are,

Number :	5.28	0.156	3	0.0428
No. of significant figures :	3	3	∞	3

The least number of significant figures here is three. Therefore, the final result is written with 3 significant figures. This is obtained by rounding off the term 51.734579 by following the rules for rounding off. The correctly reported result is 57.7.

Example 14:

Mass of a crucible is 15.02 g. The mass of this crucible and a solid substance together is 40.086 g. If the solid substance occupies a volume of 2.6 cm³. Calculate the density of the substance and report the results to the correct number of significant figures.

Solution:

Mass of the crucible + Substance = 40.086g ←3 places of decimal

Mass of crucible = 15.02g ←2 places of decimal

So, Mass of the substance = (40.086–15.02) g = 25.066g

As per rules, the difference between these two numbers should be reported up to two places of decimals. This can be done by rounding off the number. So, Correctly reported mass of the substance = 25.07g

We know. $\text{Density} = \dfrac{\text{Mass}}{\text{Volume}} = \dfrac{25.07\text{g}}{2.6\text{cm}^3}$

$$= \frac{25.07}{2.6}\text{g cm}^{-3} = 9.642\text{h cm}^{-3}$$

During division, the rule of the lowest number of significant figures is applied. So, The correctly reported density is 9.6 g cm^3; (9.6 has two significant figures as in 2.6).

Example 15:

Express the following numbers in exponential notation to three significant figures.

(a) 900027 *(b) 0.00001236*

(c) 0.000001 *(d) 406759.*

Solution:

(a) 900027 can be expressed, in the exponential notation as, 9.00027 $\times 10^5$.

Retaining only three significant figures, this number can be expressed as, 9.00×10^5.

(b) 0.00001236 can be expressed as 1.236 x 10-5. Writing this number upto only three significant figures would require rounding off the last digit. Thus, the number can be expressed as 1.24 $\times 10^{-5}$.

(c) The number 0.000001 can be expressed in the exponential notation as, 1×10^{-6}. Since, this number is to be expressed upto three significant figures, hence this number is expressed as 1.00 $\times 10^{-6}$.

(d) The number, 406759 can be expressed in the exponential form as 4.06759×10^5. To write this upto three significant figures, the pre-exponential term should be rounded off to three significant digits. Rounding this number off to the third significant figure gives 4.07×10^5.

Example 16:

State the number of significant figures in each of the following numbers.

(i) 2.653×10^4 (ii) 0.00368

(iii) 653 (iv) 0.368 (v) 0.0300

Solution:

(i) 4 (ii) 3 (iii) 3

(iv) 3 (v) 3

Example 17:

Express the following numbers to four significant figures.

(i) 5.607892 (ii) 32.392800

(iii) 1.78986×10^3 (iv) 0.007837

Solution:

(i) 5.608 (ii) 32.39

(iii) 1.790×10^3 (iv) 0.007837

Example 18:

Express the result of the following calculations to the appropriate number of significant figures.

(i) $\dfrac{3.24 \times 0.08666}{5.006}$ (ii) 0.58 + 324.65

(iii) $943 \times 0.00345 + 101$

Solution:

(i) $\dfrac{3.24 \times 0.08666}{5.006} = \dfrac{0.281}{5.006} = 0.0561$

(ii) 0.58 + 324.65 = 325.23

(iii) $943 \times 0.00345 + 101 = 3.25 + 101 = 104$

Example 19:

Express the population of India in scientific notation (use 1981 census figure).

Solution:

The population of India (1981 census figure) was 684,000,000 (Sixty eight crores eighty four lakhs). In the scientific notation, this is expressed in exponential form as, 6.84×10^8 (it has three significant figures).

Example 20:

(a) Classify the following as pure substances or mixtures.

(b) Separate the pure substance into elements and compounds, and divide the mixtures into homogeneous and heterogeneous.

(i) air	*(ii) milk*
(iii) graphite	*(iv) diamond*
(v) gasoline	*(vi) tap water*
(vii) distilled water	*(viii) oxygen*
(ix) one rupee coin	*(x) 22 carat gold*
(xi) steel	*(xii) iron*
(xiii) sodium chloride	*(xiv) iodised salt*

Solution:

The given substances are classified as follows:

Pure substances		*Mixtures*	
Elements	Compounds	Homogeneous	Heterogeneous
Graphite	Distilled water	Air	Milk
Diamond	Sodium chloride	Gasoline	Iodised salt
Oxygen		Tap water	
Iron		One rupee coin	
		22 carat gold	

Example 21:

Which of the following mixtures are homogeneous?

(i) wood	*(ii) tap water*
(iii) soil	*(iv) cloud*

Solution:

Homogeneous mixture: Tap water

Example 22:

Two substance X and Y combine to give a substance Z. The process is exothermic and Z has properties different from those of X and Y. Is the substance Z,

(i) an element

(ii) a mixture

(iii) a compound? Give explanation to support your answer.

Solution:

The substance Z is a compound. This is because;

(i) Heat is evolved during the formation of Z.

(ii) The properties Z are different from those of X and Y.

Example 23:

How will you separate the following mixtures? (suggest as many methods as you can).

(i) Salt + water *(ii) Sand + water*

(iii) Oil + water *(iv) Iron filings and saw-dust*

(v) Glass powder and sugar

Solution:

(i) Salt + Water – by crystallisation one can obtain salt in crystalline form.

– by distillation, water is obtained as distillate and the salt remains behind in the flask.

(ii) Sand + Water – by sedimentation followed by filtration.

(iii) Oil + Water – by a separating funnel (gravity method).

(iv) Iron filing + Saw-dust – iron filings can be removed by a magnet. Saw-dust remains behind.

(v) Glass powder + Sugar – dissolve the mixture in water, filter it. The glass powder remains at the filter paper and the filtrate can be concentrated to crystallise the sugar.

Example 24:

(a) When 4.2 g of NaHCO, (sodium hydrogen carbonate) is added to a solution of CH_2COOH {acetic acid) weighing 10.0 g, it is observed that 2.2 g of CO_2 is released to the atmosphere. The residue left is found to weight 12.0 g. Show that these observations are in agreement with the law of conservation of mass.

(b) If 6.3 g ofNaHCO, is added to 15.0 g of CH_2COOH solution, the residue is found to weigh 18.0 g. What is the mass of CO_2 released in the reaction?

Solution:

(a) Mass of the reactants = 4.2 g + 10.0 g = 14.2 g

Mass of the products = 2.2 g + 12.0 g = 14.2 g

Since, the total mass of the products equals the total mass of the reactants hence, the observations are in agreement with the law of conservation of mass.

(b) Let w g of CO_2 is produced in the reaction. According to the law of conservation of mass,

Total mass of the reactants = Total mass of the products

Thus, $6.3g + 15.0g = 18.0g + wg$

or $w = (15.0 + 6.3)\ g - 18.0g = 3.3g$

Example 25:

12 g of calcium carbonate was heated till no further loss in weight. The final mass was found to be 6.72 g. How much carbon dioxide was given out?

Solution:

According to the law of conservation of mass,

Calcium carbonate = Calcium oxide + Carbon dioxide

$12g = 6.72g + x$

Thus, $12\ g = 6.72\ g + x$

or $x = (12 - 6.72)$

$g = 5.28\ g.$

Example 26:

1.375 g of cupric oxide on reduction in hydrogen gas gives 1.098 g of copper. In another experiment, 1.179 g of metallic copper produced 1.476 g of copper oxide. Show that these results illustrate the law of constant (or definite) proportions.

Solution:

Experiment 1: Mass of copper oxide taken = 1.375 g

Mass of copper obtained = 1.098 g

Therefore Mass % of copper $\frac{1.098 \times 100}{1.375} = 79.86$

Experiment 2: Mass of copper oxide produced = 1.476 g

Mass of copper used = 1.179 g

Therefore, Mass % of copper $\frac{1.179 \times 100}{1.476} = 79.89$

Since, the percentage of copper in the two samples of copper oxide is the same (within ±0.5%), hence the law of definite proportion is verified.

EXERCISES

1. State the principle of measurement of a physical quantity.
2. Name the two parameters which are required to express any measurement.
3. Which of the following is more precise?
   (a) 160 cm (b) 1.6×10^2 cm
   (c) 160.0 cm
4. Name the seven base SI units.
5. How are the following quantities abbreviated?
   (a) kilogram (b) second
   (c) metre (d) millilitre
6. Which of the following is correct: 293 K or 293°K?
7. Show that the SI unit of force has the dimensions kg m s^{-2}.

8. You measure the volume of a liquid using a measuring cylinder calibrated upto 1 mL. What is the precision of this measurement?
9. Which zero in a number is significant?
10. What are exponential numbers? How can the number of significant digits be determined in an exponential number?
11. State the number of significant figures in each of the following quantities.

(a) 51.8 (b) 0.0605

(c) 9.89577 (d) 50.0

(e) 5.0 (j) 21000

(g) 16 (h) 6.023×10^9

(i) 3.6×10^9 (j) 23.40×10^5

12. State the number of significant figures in each of the following numbers,

(i) 2.653×10^4 (ii) 0.00368

(iii) 653 (iv) 0.368

(v) 0.0300.

13. Express the following numbers to four significant figures.

(i) 5.607892 (ii) 32.392800

(iii) 1.78986×10^3 (iv) 0.007837.

14. Express the population of India in scientific notation: use 1981 census figures: 684 million.
15. Add:

(i) 94.5 and 4.03,

(ii) 79.96, 14.2 and 5.7,

(iii) 2.85×10^{-3} and 4.19×10^{-2}.

16. Subtract:

(i) 0.016 from 6.17,

(ii) 14.92 from 238.5,

(iii) 0.0665 from 0.07638.

17. Multiply:

    (i) 740×250 (ii) $6.022 \times 10^{23} \times 1.00796$

    (iii) 0.08205×298.2

18. Divide:

    (i) $18.96 \div 6.939$ (ii) $48.42 \div 44.956$

    (iii) $760 + 298.2$.

19. Express the results of the following calculations to the appropriate number of significant figures

    (i) $\dfrac{3.24 \times 0.0866}{5.006}$ (ii) $0.58 + 324.65$

    (iii) $943 \times 0.00345 + 101$

20. Count the number of significant figures in the following numbers:

    (a) velocity of light: 3.0×10^8 m s^{-1}

    (b) Mass of an electron: 9.11×10^{-31} kg

    (c) 0.00368

    (d) 207.35

21. Express the number 0.0000000490 in scientific notation and count the number of significant figures.

22. Express the following upto three significant figures:

    (a) 6.5089 (b) 32.3928

    (c) 8.721×10^3 (d) Two thousand Exercises.

2

CHEMICAL CLASSIFICATION OF MATTER

INTRODUCTION

The classification of matter into solids, liquids and gases is termed physical classification of matter. 'Solid, liquid and gas are the three states of matter. Thus, matter exist in three physical states: gas, liquid and solid. Solids under ordinary conditions have definite volume -and shape, and tend to maintain these even under deforming forces.

Liquids also have definite volume but take the shape of the vessel into which they are poured. Both the liquids and solids are nearly incompressible. Gases, on the otherhand, maintain neither the volume nor shape, and completely fill the container into which they are introduced. Gases can be expanded or compressed very easily.

The three states of matter are interconvertible. This can be done by heating or cooling. Heating increases the interparticle spacing and the kinetic energy of the particles. So, a solid on heating gets converted into a liquid, and a liquid into a gas.

Many properties of solids, liquids and gases can be easily observed with the help of our sense organs. The properties which can be observed with the help of our sense organs are called macroscopic properties. The description of the behaviour of the three states of matter in terms of atomic theory is called microscopic description of matter. In chemistry, we try to explain the macroscopic behaviour of matter in terms of its microscopic description. In this chapter, we would try to explain the observable properties of different states of matter in terms of the behaviour of the constituent particles in them.

The universe is made of matter and energy. Although, matter and energy are interconvertible, but while the Matter → Energy transformation can be seen, the Energy → Matter conversion cannot. The most familiar forms of energy are heat and light.

MATTER

The matter is defined as something that occupies space, possesses mass and offers resistance to any stress applied on it. Different kinds of matter are made up of different substances. A *substance* is a definite variety of matter, all samples of which have the same properties.

Substances have two major types of properties, viz; physical and chemical. The physical properties describe any substance as it is, *e.g.*, shape, hardness or softness, melting point and boiling point etc.

The, chemical properties describe the ability of any substance to change into new substance or substances.

The physical and chemical properties form the basis of classification of matter in the following two ways.

(i) Classification of matter on the basis of physical properties: Physical classification of matter.

(ii) Classification of matter on the basis of chemical properties: Chemical classification of matter.

PHYSICAL CLASSIFICATION OF MATTER

Based upon the physical characteristics of matter, it is classified into three states, viz., solids, liquids and gases.

In solids, the particles are closely packed and bound by strong interparticle attraction. This makes solids rigid and geometrical.

In liquids, the particles are loosely packed and are bound to each other with forces weaker than those in solids. This makes liquids mobile and shapeless.

In gases, the particles are separated from each other by much greater distances, almost 10 to 100 times the size of the particle. Thus, there is virtually no force of attraction between the particles (in fact, in gases the operating forces are very weak).

As a result, the particles in gases are so loosely packed that they are free to move in any direction. This makes gases shapeless and highly compressible.

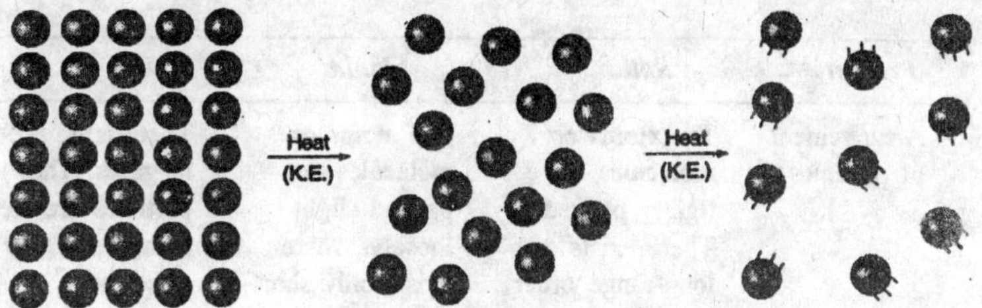

Fig. 2.1. : Effect of heating on the Interparticle spacing and the physical state of the matter.

Thus, in solids, the particles have very low kinetic energy and in gases, the particles have very large kinetic energy.

The three states of matter are interconvertible. This can be affected by heating/cooling. Heating increases the kinetic energy of the particles. As a result, a solid on heating can be converted into a liquid, and a liquid into a gas. Effect of heating on the interparticle spacing, and hence the physical state is shown in Fig. 2.1.

Solids, liquids and gases differ from each other in certain properties. A comparison of the properties of solids, liquids and gases is given below.

Table 2.1 : Comparison of the properties of solids, liquids and gases.

Property	*Solid*	*Liquid*	*Gas*
1. Shape and volume	Because of strong inter-particle forces, solids have define shape and volume	Because of slight weaker interparticle forces the liquids do not have definite shape, but possess definite volume	Because of the absence of any significant intermolecular forces, gases have neither definite shape nor definite volume
2. Compressibility and hardness	Solids are generally hard and incompressible, due to close-packed structure	Liquids are more compressible than solids, due to more empty space (void volume) in liquids	Gases are the most compressible because of large interparticle empty space

Property	*Solid*	*Liquid*	*Gas*
3. Arrangement of particles	The atoms or molecules are tightly packed. There exists a long-range order in solids.	The atoms or molecules are packed slight loosely. There exists only short-range order in liquids	There is no order in gases. The particles are free to move in any direction.
4. Diffusion	Solids do not diffuse into one another. This is due to the immobility of the particles in solids.	Liquids show slow diffusion	Gases undergo diffusion freely

CHEMICAL CLASSIFICATION OF MATTER

A complete classification of matter into fundamental groups was (and also is) a very difficult task. On the basis of chemical composition, the matter exists either as a single substance or, as a mixture containing two or more substances. A sample containing only one substance is called a pure substance. A pure substance shows the following characteristics:

(i) a pure substance contains only one kind of atoms or molecules.

(ii) a pure substance is perfectly homogeneous.

(iii) a pure substance has a definite composition which does not change with time.

Pure substances can be further classified into

(a) Elements, (b) Compounds

Samples containing more than one substances are called mixtures. There are two types of mixtures.

(a) Homogeneous mixtures, (b) Heterogeneous mixtures

A *homogeneous* material is that substance which is perfectly uniform in its composition throughout. A heterogeneous material is the one which has different composition, and different properties in different parts of the sample.

A broad classification of matter is summarised in Fig. 2.2.

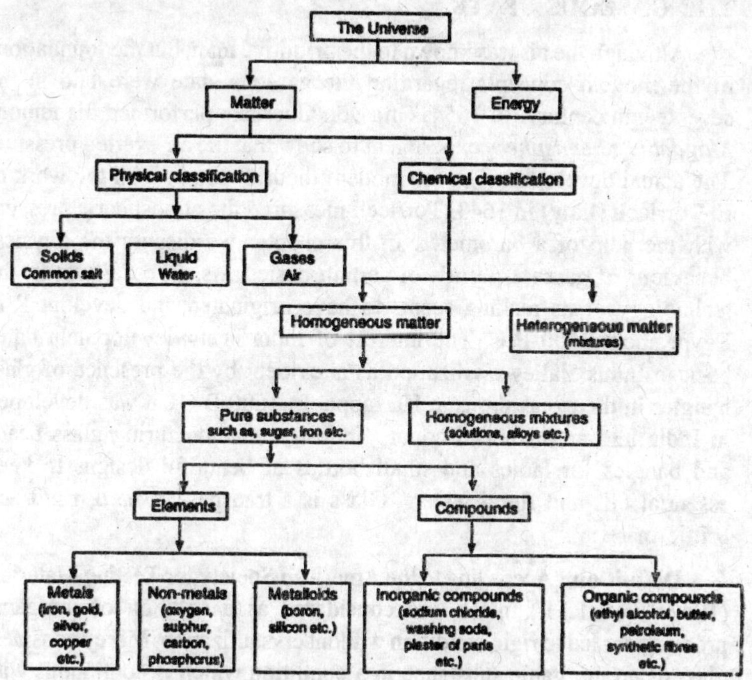

Fig. 2.2 : Classification of matter.

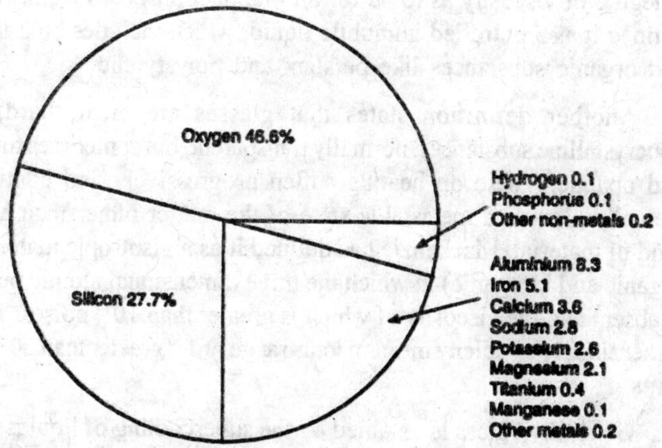

Fig. 2.3 : Percentage distribution of abundant elements in the earth's crust.

THE GASEOUS STATE

Although the air was known to the primitive man, but the foundations of the modem concepts regarding the gaseous state were laid in the seventeenth century. In 1654, Otto von Guericke performed his famous *Magdeburg hemisphere* experiment to show that the air exerted pressure. The actual development of the modem theories began with the work of E. Torriceli (Italy) in 1643. Torriceli measured the atmospheric pressure with the help of a barometer. In this chapter we discuss the physical behaviour of gases in term's of certain basic laws. In other words, the technology of its making seems to have originated and developed' in Egypt about 3000 B.C. The interest of India in its development dates back to Indus Valley civilization as is evident by the presence of glass bangles in the excavations at Harrappa. In 1600 B.C., it was developed in India into an advanced cottage industry, manufacturing glass beads and bangles for ladies and small bottles of beautiful designs to keep essential oils and floral waters. Glass is a transparent, hard, rigid and brittle material.

Definition : According to the American Society for Testing Materials (1945) (A.S.T.M.), glass was considered as a solution of inorganic products cooled to rigid condition without crystallization. Morey considers glass as an inorganic substance in a condition which is continuous with and analogous to the liquid state of that substance, but which, as a result of reversible change in viscosity during cooling, has attained so high a degree of viscosity as to be for all practical purposes rigid. Holloway defined it as a putrefied immobile liquid, which includes both inorganic and organic substances like perspex and polystyrene.

Another definition states that glasses are clear, hard, brittle, noncrystalline substances, normally transparent, but sometimes translucent and opaque. These on heating soften progressively and continuously. Actually glass is a metastable state of the matter rather than a specific kind of material. Mackenzie has defined it as an isotropic material (both organic and inorganic) in which the three dimensional atomic periodicity is absent and the viscosity of which is greater than 10^{14} poises. The three dimensional periodicity means a long-range order greater than 20 angstrom units.

Generally, glass is obtained by the supercooling of liquids, but they are also formed by the condensation of vapour. The properties of the glass formed under the two systems need not be the same. In the case of ethyl alcohol, the specific heat and the glass transition temperature are similar, both in the case of glass prepared from the vapour and the supercooled liquid.

TRANSFORMATION TEMPERATURE

The temperature at which an abrupt change in the coefficient of expansion takes place in a glass heated at about 4°C per minute is known as the transformation point Tg. It is a characteristic temperature for a given composition. Since the molecular vibrations just begin at this temperature, it is also known as the lower critical point. Actually, it is a transition region, above which it is fluid or rubbery and below which it is glass. This is because of the interaction between various chains. According to modem point of view, it is a temperature at which configuration becomes comparable, with the rate of approach to equilibrium with the time required to make an observation. The corresponding viscosity is $10^{13} - 10^{14}$ poises.

The transition region in organic glasses is characterized by changes in derivative functions, such as specific heat and coefficient of expansion. Any molecular theory of their properties leads to molecular interpretation of the glass transition, *e.g.*, ΔC_p is close to 11.2 $J.K.^7$ per bead where bead is identified as the smallest unit in polymer chain capable of free rotation.

The glass transition temperature is associated with the fractional free volume having a constant value 0.025. Attempts to correlate Tg to polymer structure have focussed on either molecular cohesion or chain stiffness or both. Gibbs and coworkers predicted a second order transition in polymers at a temperature T_2 which is about 60°K below Tg and is nearly equal to the temperature at which the axtrapolated viscosity of the liquid tends to infinity. The molecular motions are found to occur in glass polymers, but in most of the cases it is not clear as to what they are. Therefore, it is safe to assume that glass transformation is associated with large-scale movements.

PHYSICAL CHARACTERISTICS OF GASES

All gases show some common characteristics. These are described below.

(i) Gases maintain neither the volume nor shape, and completely fill the container in which they are introduced.

(ii) Gases expand appreciably on heating.

(iii) Gases are highly compressible, that is, when the pressure is increased, the volume of a gas decreases.

(iv) Gases diffuse rapidly.

(v) All gases, except a few, are colourless. The following gases show characteristic colours, such as,

Fluorine	Greenish-yellow
Chlorine	Greenish-yellow
Bromine	Reddish-brown
Iodine	Violet
Nitrogen dioxide	Reddish-brown

(vi) Gases exert pressure equally in all the directions.

MEASURABLE PROPERTIES OF GASES

The behaviour of gases can be described in terms of certain parameters. These are described below.

Mass and Amount

Mass of a certain amount of a gas can be obtained by direct weighing. TB is can be expressed in any mass unit. The most common unit of mass is gram (g) or kilogram (kg).

In chemistry, according to the IUPAC recommendations, the amount of a substance is expressed in terms of the number of moles (n). The number of moles can be obtained from the mass of the gas by using the relationship,

$$\text{No. of moles } (n) = \frac{\text{Mass of the gas in gram } (m)}{\text{Molar mass of the gas in gram per mole } (M)} \quad ...(1)$$

Volume (V)

Gases fill the container in which they are placed. So the volume of gas is equal to that of its container. Volumes can be expressed in litre (L), millilitre (mL), or cubic centimetre (cm^3). The SI unit of volume is cubic metre (m^3). In common practice, however smaller units such as cubic decimetre (dm^3) and cubic centimetre (cm^3) are more frequently employed. These are related to each other as,

$1\ m^3 = 10^3\ dm^3 = 10^6\ cm^3$

The commonly used unit, litre (L) has now been redefined so that,

$1\ L = 1\ dm^3$

and $1\ mL = 1\ cm^3$

The volume (V) of any gas depends upon its amount, temperature and pressure. Mathematically, volume of a gas is a function of the amount, temperature, and pressure, *i.e.*,

V = f (amount, temperature, pressure)

or V = f(n, T, P)

Pressure (P)

Pressure is force per unit area. Gases exert pressure uniformally in all the directions. In laboratory, the atmospheric pressure is measured with the help of a barometer, in terms of height of the mercury column in the barometer. The atmospheric pressure (P) is expressed as,

$$P = h \rho g \quad ...(2)$$

where h is the height of mercury column in the barometer

ρ is the density of mercury

g is the Laws of Chemical Combination due to gravity

A standard pressure of one atmosphere (1 atm) is defined as the pressure exerted by exactly 76 cm of mercury column at 0°C, (density of Hg = 13.5951 g cm^3) and a standard gravity of 981 cm s^{-2}. The atmosphere unit is not a SI unit.

The SI unit of pressure is Pascal (Pa). It is defined as the pressure exerted when a force of 1 newton (N) acts upon an area equal to 1 m^2. This unit is related to atm unit by the relationship,

$1\ atm = 1.01325 \times 10^5\ Pa$

Another non-SI unit generally employed in reporting pressure values is termed torr. One torr is equal to the pressure exerted by 1 mm of mercury column at 0°C, and a gravity of 981cm s^{-2}. Thus,

1 atm = 760 Torr

Thus, 1 atm = 76 cm Hg = 760 mm Hg = 760 Torr = 1.01325 x 105 Pa

Measurement of Pressure

The pressure exerted by a gas is measured by a manometer. It consists of a U-tube partly filled with a non-volatile liquid: generally mercury is used as the manometric liquid. Two types of manometer are commonly employed. These are,

Open-end manometer. In this manometer, one end of the U-tube is connected to the vessel containing the gas, and the other end is open to the atmosphere as shown is Fig. 2.4(a). The difference in the levels of mercury (h) in two limbs of the manometer gives the pressure head (Δp): Δp may be negative or positive. When the level in the limb open to the atmosphere is higher then, Δp (or h) is positive and the pressure of the gas is higher than the atmospheric pressure. Thus,

$$P_{gas} = P_{atm} + \Delta P = P_{atm} + h \rho g \quad ...(3)$$

where h is the difference in the levels of mercury in the two limbs, ρ is the density of mercury and g is the Laws of Chemical Combination due to gravity (please take the proper units of these quantities while solving numerical problems).

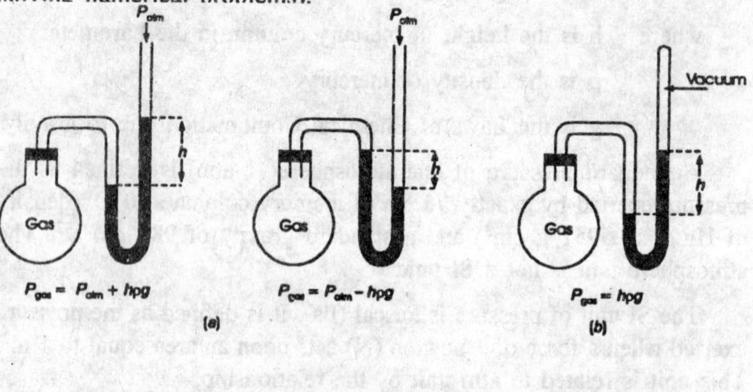

(a) (b)

Fig. 2.4 : (a) An open-end manometer, (b) A closed-end manometer.

If the mercury level in the limb open to the atmosphere is lower, then Ap is negative, and the pressure of the gas is lower than the atmospheric pressure. Thus,

$$P_{gas} = P_{atm} - \Delta P = P_{atm} - h \rho g \quad ...(4)$$

Closed-end manometer. In this type of manometer, one end of the U-tube is sealed (closed) after evacuating, while the open end is attached to the vessel containing the gas (Fig. 2.4b). The gas pressure is then given in terms of the difference between the level of mercury columns in the two limbs, through the relationship,

$$p_{gas} = h\,\rho\,g \qquad ...(5)$$

Temperature (T)

Temperature is a measure of the degree of hotness. In gases, it is a measure of kinetic energy possessed by the molecules. The instrument which is commonly used to measure the temperature of any substance is called thermometer. The temperature scale mostly used in the scientific work is the Celsius scale (formerly called centigrade scale). The freezing point of water corresponds to zero (0°C) and the boiling point of water under 1 atm pressure is 100°C on this scale. The value of 0°C has been accepted as the standard temperature as per international agreement. Temperatures are also expressed on the Kelvin scale, whose zero corresponds to –273.15°C. Thus,

Temperature on the Kelvin scale = 273.15 + Temperature on the Celsius scale

or $$T(K) = 273.15 + t°C \qquad ...(6)$$

Normal Temperature and Pressure. The temperature of 0°C (or 273.15 K) and pressure is equal to 1 atm (= 760 Torr = 1.01325×10^5 Pa) are called *normal temperature and pressure (or NTP) or standard temperature and pressure* (STP) respectively.

The NTP (or STP) conditions are generally used for comparing the behaviour of gases.

GAS LAWS

The macroscopic behaviour of gases can be described by certain generalizations. These generalized observations are called gas laws. These are described below.

Boyle's Law: (Pressure-Volume Relationship)

Robert Boyle (1622) studied the effect of pressure on the volume of gases when temperature is held constant. This relationship known as Boyle's law is stated as follows.

"At constant temperature, the volume of a certain amount of a gas is inversely proportional to the "pressure exerted on it."

This relationship has been found to be true for almost all gases under ordinary conditions. If V is the volume of a certain amount of gas under-pressure P at a fixed temperature, then

$$V \propto \frac{1}{P}$$

$$PV = \text{constant} \qquad ...(7)$$

The value of the 'constant' in Eq. 2.7 depends upon the amount and temperature of the gas. Consider a given amount of a gas occupying a volume V_1 under pressure P_1. Then, from Boyle's law,

$$P_1 V_1 = \text{constant}$$

Now, if temperature is kept constant, and pressure is changed to P_2 so that the volume changes to V_2. Then, from the Boyle's law,

$$P_1 V_1 = \text{constant}$$

Now, since temperature and the amount of the gas remains unchanged hence,

$$P_1V_1 = P_2 V_2$$

where P_1 is the original pressure, P_2 is the new pressure

V_1 is the original pressure, V_2 is the new volume

Thus, if the same amount of a gas is subject to a change of pressure at constant temperature, the product of each pressure - volume will be equal. Then, according to the Boyle's law,

$$P_1 V_1 = P_2 V_2 = P_3 V_3 ..... = \text{constant} \qquad ...(8)$$

Graphical Representation of Boyle's Law

The Boyle's law relationship is shown graphically as the plots of,

Volume vs Pressure (V vs P)

or Volume vs $\frac{1}{\text{Pressure}}$ $\left(V \text{ vs } \frac{1}{P}\right)$

Pressure × Volume vs Pressure (P V vs P)

These plots are shown in Fig. 2.3.

(a) *Volume–pressure graph :* When the volume (V) of a certain amount of a gas (at any fixed temperature) is plotted against pressure (P), a hyperbolic curve is obtained. The hyperbolic curve obtained when volume (V) of a gas is plotted against pressure (P) at any constant temperature is called volume–pressure or pressure–volume isotherm, (V–P isotherm, or P–V isotherm).

The volume–pressure isotherms for a gas at three different temperatures are shown in Fig. 2.5 (a).

(b) Volume–l/Pressure graph. When volume (V) of a certain amount of gas (at any constant temperature) is plotted against reciprocal of pressure (1/P), a straight line plot passing through the origin is obtained. Thus, V–1/P (or V–P^{-1}) isotherms are straight lines passing through the origin. The V–1/P isotherms at three different temperatures are shown in Fig. 2.5(b). As you see in the figure, the slope of straight line plot increases as the temperature of the isotherm increases.

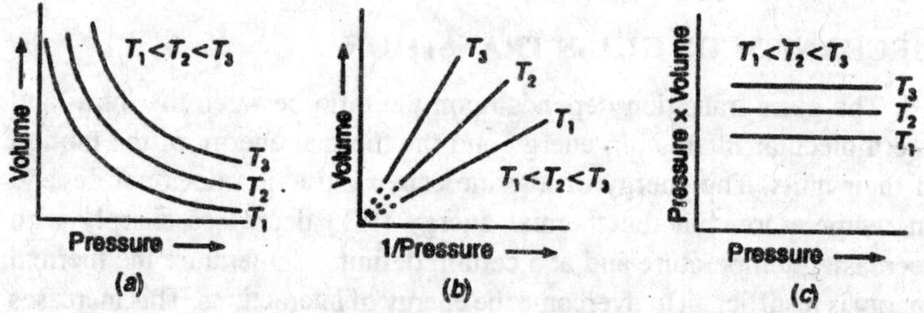

Fig. 2.5 : The graphical presentation of the Boyle's law: (a) V vs P (b) V vs 1/P and (c) PV vs P.

(c) *Pressure × Volume us pressure graph :* When the pressure × volume product (PV) for a certain amount of gas is plotted against pressure (P) at any constant temperature, the isotherm is a straight line parallel to the pressure-axis. Thus, the PV at any constant temperature, the isotherm pressures (At near atmospheric pressures) Thus, PV vs P isotherms are straight lines parallel to the pressure axis. The PV vs P isotherms for a gas at three different temperatures are shown in Fig. 2.5(c). At any pressure, the value of the product PV increases with an increases in temperature.

RELAXATION

The term relaxation expresses the establishment of a static equilibrium in a physical system. In a liquid continuous regrouping of molecular clusters appear under the influence of thermal mobility. Boltzmann's law suggests a method of working out a probability of the process. The reciprocal of this probability of regrouping gives the relaxation time, which is of the order of $10^{-4} - 10^{-6}$ sec in the case of polymers as compared to 10^{-10} sec in the ordinary liquids. Thus, polymer chains possess low mobility and higher relaxation time. The process in which equilibrium is established with time is known as relaxation process.

The curve between temperature and deformation. This consists of three positions, namely glassy state, rubber like state and viscofluid state. The transition from the glassy state to rubber-like state involves a change in the nature of super molecular structure but the phase remains the same as there is no heat effect. The difference between glassy state and the rubber-like state is in the mobility of their macromolecules and super molecular structure. Thus, glass transition is not a phase transition.

MECHANISM OF GLASS TRANSITION

The glass transition depends upon the ratio between the intra- and intermolecular interaction energy and the thermal energy of the motion of their units. The energy of intermolecular attraction does not depend on temperature, but the thermal energy (KT) decreases sharply with decreasing temperature and at a certain definite temperature the thermal energy is insufficient to overcome the energy of interactions. This increases the viscosity of the polymer and decreases the intensity of thermal motion of its units.

METHODS OF DETERMINING T_g

The transition of a polymer from the rubber-like to glass state is accompanied by a gradual change in its physical properties and T_g can be determined by studying the variation of these properties with temperature. The most popular methods are:

(1) specific volume (dialatometric method)

(2) heat capacity

(3) modulus of elasticity

(4) deformation

The difference between a glass and the corresponding liquid or supercooled liquid may be demonstrated by the isobaric change in volume with temperature for an imaginary polymer. The polymer melt on cooling can form (a) a crystal or (b) a glass depending upon experimental conditions. If the substance crystallizes, *i.e.*, follows the path A, B, C, D, it is observed that as the liquid cools from any temperature T, to T_f, there is jump in the volume at T_f, the freezing point. Normally there is marked decrease in volume but in some cases like water, it increases. The coefficient of expansion of the liquid and of the crystal are constant in the vicinity of T_m. As the liquid is cooled slowly, its molecules arrange themselves into a definite pattern forming a crystal. In some of the liquids the forces of attraction (covalent or hydrogen bonds) between the molecules are so strong even above the freezing point, that they cannot freely move and do not provide a suitable crystal nucleus, the crystallization is prevented. When the liquid is rapidly cooled beyond its freezing point T_f the molecules do not have the time to organize themselves all over the liquid and hence, it is supercooled. The individual atoms or molecules are in an ordered array over a short range, but there is no long-range order.

They are indeed in structure, very much like a snapshot of a liquid and are called supercooled liquid. Molecular movement in such liquids is minute and the attempt to describe the flow properties in terms of viscosity to the rigid glass is difficult. Below Tg*l* the dependence is again linear.

Te is the start of the solidification and Tg*l* is the termination of the solidification. The temperature at the intersection of the extrapolated curves Tg is known as glass temperature. The molecules in the glassy state oscillate about their equilibrium position and their interchange of places either does not occur at all or only very rarely.

The decrease in the thermal energy of the molecules with the fall in temperature decreases their ability to break down strong covalent bonds, when most of the motion is frozen at a temperature Tg*l*, bearing the liquid in a random and disordered state, a solid state results which is known as glassy state. One may say that the glass represent the metastable state of order of a liquid which is frozen in the range Te & Tg*l* and preserved at the lower temperature rather than a specific kind of material.

For the supercooled liquid in the range Tg and Te at a given rate of cooling (glass 1 A 2) the equilibrium distribution corresponding to each temperature is reached within the duration of the measurement, since the substance is in thermodynamical equilibrium with reference to arrangement and distribution of molecules.

When a liquid on cooling changes to crystal, there is an evolution of latent heat associated with the abrupt change in the disorder, but there is no evolution of latent heat during the liquid to glass transformation. This suggests that the molecular arrangement in glass is similar to that in liquid. The glassy state is obtained by the inhibition of a kinetic process rather than a second order transition.

STRUCTURE

The structural units in glass are randomly linked and the regularity in the lattice has been found to extend to 5–10° A only. So it is evident that it has only a short-range order.

Silicates are known to exist in three crystalline and one amorphous form as shown below:

$$\text{Quartz} \xrightarrow{870^\circ} \text{Tridymite} \xrightarrow{1470^\circ} \text{Crystabollite} \xrightarrow{1710^\circ} \text{glass}$$

Silicon atoms are bound to four oxygen atoms forming a regular tetrahedron. The distance between the centres of the silica and oxygen ions is 1.6° A. In crystoballite the structure consist of a set of interlocking rings each with six atoms, three silicon and three oxygen held together by covalent bonds forming a giant molecule like that of carbon in diamond. All the three varieties soften at 1600° and melt to a liquid at 1700°. The silica crystal can melt only by breaking of the bonds. In the liquid a large number of Si-O bonds have broken to allow the parts to move about.

However, the breaking is not irrevocable; Si-O bonds are continuously forming and breaking in the melt so that most of the time silicon is attached to four oxygen atoms and oxygen to silicon atoms. When the melt is cooled rapidly, the particular bond pattern present at that instant is frozen to a liquid of very high viscosity Instead of a regular pattern of six membered rings, the tetrahedron is distorted resulting in jumbled rings of different shapes and sizes. The positional disorder existing within each distorted tetrahedron prevents the prediction and assignment of fixed position to the ions.

The resulting glass is still a giant molecule but a disordered one. It is debarred from changing into a regular structure since that would involve the breaking and reforming of strong covalent bonds. The strong directional covalent bonds, present in the liquid, form wrong connections leading to glass. Although, a short-range order is followed they form rings with the wrong number of elements. Moreover, the atoms in glass are at rest and there is a fairly constant distance between each atom and its -neighbour. The large interatomic distance do not have the same regularity as in the crystals. Therefore, the glasses have partially ordered material as is evident from two or more diffused haloes in the X-ray photographs. In view of very strong covalent bonds and very high viscosity of the liquids, the probability of the atoms to break their connections and to reform in the right direction is very low. Therefore, they form a disordered arrangement with strong interactions, resulting in glass. Since different energies will be required to breakdown the different meshes of structure, there is a continuous softening range instead of abrupt points.

Warren ruled out the possibility of the existence of molecules in the glass and found very good confirmation in the calculated and observed-maximas. It was very convenient to describe the change in viscosity on cooling to an association or polymerization process.

Zachiarasen deduced the following condition to be followed by an oxide glass of the composition $A_m O_n$: (1) No oxygen atom can be bound to more than two A atoms, (2) The coordination of oxygen around A must be low. (3) The oxygen polyhedra must not share edges or faces but only comets, and (4) At least three comers of the oxygen polyhedra must be shared.

There are two types of substances that follow the above conditions:

Those substances, that have directed covalent bonds, crystallize in an open structure, *e.g.*, SiO_2, B_2O_3, GeO_2, P_2O_2, As_2O_5, V_2O_5, Sb_2O_3, Sb_2O_5, As_2O_3, ClO_5, Ta_2C_5.

Substances whose liquids consist of long molecules which become entangled on cooling and require breaking of many covalent bonds for rearrangement, *e.g.*, plastic sulphur, organic glasses.

ORGANIC GLASSES

It is perhaps not always recognized that glass formation is a property shared by a very large class of compounds. These are polyphosphate glasses, those formed by certain elements in the polymeric form, *e.g.*, S, Se, those formed by hydrogen bonded materials like alcohols and organic polymers.

Lourenco obtained a highly polymerized viscous undistillable residue by polymerizing ethylene glycol in presence of ethylene dihalide, but Staudinger was the first who suggested chain formulas for organic polymers. The polymeric glasses comprises a class of plastic materials of long chain organic molecules having an amorphous structure. The polymer chains consist of single repeating monomer units joined together by covalent bonds. Polanyi and Herzon suggested that no unit cell of crystal can be smaller than a single molecule. In 1929 Staudinger clarified the confusion regarding the structure of polymers.

However, certain plastics are also used as copolymers, *i.e.*, polymers containing more than one monomer unit, *e.g.*, styreneacrylonitrite copolymers of vinyl chloride, vinylacetate and nylon, synthesized from adipic acid hexamethylene diamine with an elementary unit—$NHCO(CH_2)_4CONH(CH_2)_6NH$—of a length of 17°A, which on melting at 260°C gives a true glass. Carothers and associates made an effort to link physical properties, *e.g.*, viscosity, double refraction, rubber elasticity, etc., with molecular weight and shapes of these molecules. Flory has worked out several methods to determine the various physical properties of the polymers. Zachariasen propagated the theory of random continuous network for polymers.

In the case of linear polymers each chain molecule retains only two functional end groups during the process of polymerization irrespective of its size. The main chemical type of polymer glass may be represented by polystyrene, polymethyl methacrylate, polyvinylchloride, and polycarbonates, etc. Thus, polystyrene normally contains a small quantity of lubricant which reduces the glass transformation temperature by 10°C and improves its moulding properties.

In nonlinear polymers the number of functional groups increase for each molecule with its size or us molecular weight, *e.g.*, formation of various sulphur, where S_8 ring opens and produces a chain between two molecules by the pairing of electrons.

Polymethyl methacrylate is a glass below 90°C. The mixture of polymer and monomer, or of polymer, monomer and the diluent, transform to a glass. During the process of polymerization, the percentage of polymer reaches a certain well defined high value. Even the monomer molecules are frozen up in their position in the glassy state.

Recently, the organosilicone compounds, because of their greater stability, have become important. The $Si\text{-}CH_3$ or $Si\text{-}C_2H_5$ bonds have greater oxidation resistance than the corresponding C-C bonds. Depending upon the molecular weight, chain length, ring structure and the degree of three-dimensional polymerization, the methyl or ethylsilicones can have very different properties ranging from fluid to viscous oil, resins and elastomers. The products obtained depend on the methods of preparation.

Warren, Bagchi, Alexander and Miller, have discussed and given upto date account of the application of X-ray to glassy materials. Debye, Station and Heikens have used the X-ray scattering method for determining the void spaces in polymers and glasses. Interesting relationships have been observed between the heterogeneity of the glass and mechanical properties of fibres. Several workers have also applied NMR, ultrasonic and diffusion techniques for the above purposes.

The constitution of an organic polymer is guided by the following parameters:

1. The configuration describes the atomic structure of the molecule according to the chemical formula. The configuration is the same in solution, in the glassy state or the crystalline state.
2. The conformation is a dynamic parameter, *e.g.*, the rotation of the segments about a single bond. They may have different number of elementary units per turn of the helix but in every case the number of possible conformation will be greater in liquid or in the glassy state than in the crystal.
3. The molecular packing or chain packing refers to the lateral organization of the chains in the crystal. Since most of the polymers even as crystals are only poorly defined, this parameter covers different types of disorders and lattice imperfections.

Du Pont have reported new polymers called ionomers by crosslinking chains of polyethylene through the ionic forces acting between (–COO) groups and Mg^{+2} or Zn^{+2} as cations. Rees and Vaughan have been successful to introduce strong ionic forces between the chains through their carboxylic groups.

The ionomers promise a rare insight into the effect of gradual transition of Van der Waals forces and hydrogen bonding into the three-dimensional ionic network. The ionomer resins are solids with a high vibrational entropy, introduced by the electrical fields, screening demands and nucleating effects of cations which interfere with the éstablishment of long-range order. All these facts result in the following properties:

1. The stiffness increases with the acid level and the concentration of cations.
2. The ionomer resins exhibit unusual stability in high voltage application.
3. The ionomer resins have unusual elasticity and toughness.
4. The ionomers, because of association in liquid state resulting in high viscosity, behave like high molecular weight thermoplastics than like cross-linked polymers.
5. Compositions containing dibasic acids in conjunction with di- or trivalent ions resemble cross-linked olefins in the behaviour of the melt.
6. The temperature coefficient of viscosity is very sensitive to the strength of bonding forces.

Covalent C-C bond in cross-linked polymers present flow at temperatures at which the polymer is thermally stable. Ionic bonds and resulting association in ionomers makes the viscosity of the ionomer very sensitive to temperature giving activation energy of the order of 20 kcal/mole.

Hydrogen bonding through carboxylic acid molecules makes the viscosity less sensitive to temperature where the activation energy is 14 kcal/mole. Van der Waals forces in the polymers show the lowest temperature coefficient, *i.e.*, 7-19 kcal/mole.

The intermolecular binding forces determine its solubility in organic solvents and permeability for oils, lonomers have extremely low permeability for oil and are insoluble in common solvents. Special mixtures, *e.g.*, N, N-dimethylacetamide, $CH_3CON(CH_3)$; added to decalin can dissolve ionomer.

Rose and Block have synthesized some inorganic polymers by refluxing dibutylphosphoric acid with zinc acetate. The above polymer remains flexible down to –80°C. It is soluble in chloroform and a polymer of molecular weight 10,000, melts at 150°C and forms a glass on cooling.

INORGANIC GLASSES

Obsidion, a mineral found in nature, is a dark reddish brown, translucent glass, formed when molten rock is rapidly cooled. These inorganic glasses are the products of fusion of inorganic salts cooled to a rigid mass without crystallization. Winter observed that oxygen, sulphur, selinium, and telurium elements of group VI of the periodic table form glasses with other fifteen elements of group III, IV, and V. These group VI elements are found to form binary glasses and are self vitrifying. A discussion about these is beyond the scope of the present work and interested readers are advised to read the original work. These include various oxides, chlorides, nitrates, sulphates, thiosulphate, alums and other hydrated salts.

DEVITRIFICATION

The molten mixtures of alkali silicates, on cooling slowly, try to crystallize out into one or more of the compounds at their freezing point and the whole mass loses its glassy nature. This phenomenon is known as the devirtrification of glass. All glasses except those having very high viscosity in the molten state try to devitrify. Reamer produced some polycrystalline ceramics by heating glass but it was actually Stookey who systematically studied this subject and actually produced ceramic by heating glass under controlled condition. The small devitrified glass crystal slowly grow into a strong, opaque, white mass, known as porcelain. They have a much higher mechanical strength and markedly improved electrical insulation properties than ordinary glass.

VISCOSITY TEMPERATURE CHARACTERISTICS

Glass on heating softens continuously and progressively to fluid due to continuous fall in viscosity.

Gehlhoff and Thomas found a correlation between the change in viscosity and its composition. Therefore, the relationship between viscosity and temperature is a very important parameter for the glass. Important work in this direction is carried out by Field, Royster, Herty and McCaffery. They have made a thorough study of change in viscosity with temperature and with time. Lillie has produced the most accurate results using modified Margules method. The complete viscosity temperature curve is given in Fig. 3.27. It would be of interest to define the viscosity temperature relationship in terms of certain characteristic temperature at which the glass attains a particular value for viscosity.

Certain temperatures are only of historical importance, *e.g.*, deformation temperature, cohesion temperature, flow temperature, softening point, etc., while others like working point, Mg point, annealing point, strain point and transformation point are very important in the manufacture of glasses.

Working Point

At the usual melting temperatures glass has the same viscosity as glycerine at 0°C, *i.e.*, 10^2 poise. For giving a shape to the glass, it should remain in the fluid condition over a wide range of temperature. Therefore, the glass is removed from the furnace at a viscosity of $10^4 - 10^5$, poise so that it can be given proper shape. The corresponding temperature is known as working or flow temperature.

Softening Point

Littleton defines it as the temperature at which a uniform rod of 0.5 to 1.00 mm diameter and 22.9 cm long elongates under its own weight at a rate of 1 mm per minutê under standard conditions. The viscosity at this temperature is $10^6 - 10^7$ poises.

Mg Point

The temperature at which the viscosity flow exactly counteracts the thermal expansion during the measurement is known as dilatometric or Mg point. According to the recommendations of the Society of Glass Technology it is the maximum point reached on the complete thermal expansion curve for the glass, *i.e.*, the point normally corresponding with the softening temperature.

The viscosity at this temperatures is $10^{11} - 10^{12}$ poises. It also corresponds to the extreme upper limit of the annealing range and is denoted by A or the upper critical point designated by C_r. This temperature also depends on the thermal history of the glass.

Annealing Point

Glass on being cooled, without special heat treatment, develops unequal strained condition leading to a tendency to break. To prevent this strain, exact controlled temperature is maintained during a short interval known as annealing range.

The temperature at which the internal stress in the glass is relieved within fifteen minutes, is known as annealing point. The viscosity corresponding to this temperature is 10^{13-14} poises.

Strain Point

Littleton defines the lower limit of the annealing point as the strain point. This is also the highest temperature from which the glass can be rapidly cooled without introducing serious internal stresses. The glass can be commercially annealed within sixteen hours. The viscosity at this temperature is 10^{14-16} poises.

GLASS TRANSITION TEMPERATURE AND MOLECULAR MASS

Glass transition increases rapidly with the increasing mass and finally reaches a constant value. The molecular mass at which T_g becomes constant depends on its chain flexibility. For isobutylene T_g becomes constant at M equal to 1000 while polystyrene M should be more than 15,000.

Low molecular weight polymers exist only in two states, *i.e.*, the glass and the liquid and the mass transition temperature coincide with the flow temperature. But, with increasing molecular weight at a certain molecular mass the transition temperature splits into T_g and T_f, thus, introducing a rubber-like state. With further increase in mass, T_g becomes constant but T_f continues to increase. Thus, $T_f - T_g$ gives the temperature range of rubber-like state.

It has been experienced that low molecular mass polymers exist as glassy or liquid where their T_g and flow temperature coincide. With higher molecular mass $T_g - T_g$ is great showing the increase in rubber range. The molecular mass at which the transition temperature splits into T_g and T_f depends on the chain flexibility. The more rigid the chain, higher will be the molecular mass at which the T_g splits. Thus, in polyisobutylene rubber-like strain appears at molecular mass 1000 while in polystyrene, it appears at M 40000.

Polymers with less flexible chains have very high T_g and they do not display rubber-like structure even at higher temperatures. In such case, we talk of softening with a range of temperature. T_f of amorphous polymers increases with increasing molecular mass.

These fact go a long way in the synthesis of polymer with desired properties.

In the case of polydisperse polymers, polymer fractions of different molecular mass pass into visco-fluid state at different temperatures.

GLASS TRANSITION TEMPERATURE AND CHEMICAL CONSTITUTION OF POLYMERS

Polymers with polar molecules give easily glassy state because of high interaction energy. The links between polar groups of one chain with polar group of the other neighbouring chain are stronger and do not break unless the thermal motion is very intense. Thus, they form local cross-link points in the polymer which are not constant in time and go on breaking and forming a new. The average life time of groups in the linked state increases with the decrease in temperature and becomes appreciable near the glass transition temperature.

The formation of crosslinks limits their thermal motion. This results in making the entire system more rigid, even in presence of small number of crosslinks, forming a structure of fixed random arrangement of polymer molecules.

The polymer acquires the properties of a solid below T_g. It has also been observed that in glassy polymers most of the hydroxyl groups takes part in hydrogen bonding.

Branched-chain and linear aliphatic resins provide low crosslinks and higher chain flexibility. One group on a given molecule with more than one reactive site on the adjacent molecule tends towards cyclization process rather than crosslinking. The effect of the types of linkages and T in the case of epoxy resin.

FORCED ELASTICITY

Low molecular glasses like resin and silicate glasses do not respond to large stresses. Macromolecular glasses often becomes brittle at temperatures much below T_g. The ability of glassy polymers to deform considerably without breaking is responsible for their wide use in engineering. The large deformation of glassy polymers was at one time attributed to flow process called cold flow, however, mutual displacement of macromolecules is not probable.

Methyl methacry late under a strain at temperature below T_g was found to recover its initial shape and size after heating above T_g.

This recoverable nature of large deformation of high molecular glasses suggests that the regularities of rubber-like state apply to the glass of state as well, showing that the glass transition is not a phase transition and a polymer possesses the same structure at temperatures above and below T_g. The relaxation time of polymer in glassy state is very great.

Equation 1 gives a relation between relaxation time and temperature

$$\tau = \tau_o C^{\Delta u-\alpha s/kT}$$

where τ is the relaxation time, ΔU is the activation energy, a is the stress and a is a constant.

At small stresses, stress value does not affect relaxation time but at larger stresses τ becomes smaller and highly elastic strain appears. It can also be seen that after the removal of deforming stress, elastic strains do not disappear at temperatures below T_g.

STRESS STRAIN DEPENDENCE

Glassy polymers like polystyrene, poly (methyl methacrylate), poly (vinyl chloride), etc. on deformation begins to extend non-uniformly when a particular stress is reached. One part of specimen becomes thinner than the rest of the mass and a neck is formed. Glassy polymers like cellulose acetate, cellulose nitrate do not form neck upon forced elastic deformation. The stress at which the total rate of deformation becomes equal to force elastic deformation is called ultimate forced elasticity.

The ultimate forced elasticity has been found to rise with increasing straining rate indicating its relaxation nature.

EFFECT OF MOLECULAR MASS

The molecular mass affects the brittle strength and T_g of the polymer. T increases with molecular mass of a polymer up to n = 200 (n is degree of polymerization) and the brittle strength up to n = 600. With increasing n both T_g and σ_{br} increases up to n = 200. With further increase of n, T_g remains constant, but σ_{br} continues to increase and T_{br} decreases.

Temperature T_f is the viscofluid region, between T_f and T_g is the rubber-like state region, between T_g and T_f is the forced elasticity region and below T_{br} the polymer is in the brittle state.

GLASS TRANSITION BY CHROMATOGRAPHY

Partition of solute molecules between gas and stationary phases, expressed in gas chromatography as a retention volume, is a characteristic of the system investigated. A plot of the logarithm of the retention volume versus reciprocal of absolute temperature should be a straight line whose slope is directly related to the corresponding enthalpy, heat of solution in gas-liquid chromatography (GLC) and heat of adsorption in gas—solid chromatography (GSC). Polymer stationary phases deviate from such linear behavior at their glass transition temperature (T_g).

It has been shown that a change of retention mechanism occurs at T_g. While below T_g (region I) retention proceeds exclusively by surface adsorption, increasing molecular mobility at and above T_g (region II) is accompanied by an increase in retention volume due to bulk sorption. At even higher temperatures (region III), equilibrium conditions are restored, and bulk retention volumes decrease with increasing temperature. Because of the high sensitivity of GC detectors, measurements can be made with very low concentrations of solutes at a level where they do not alter the structure of the polymer under study.

Due to the occurrence of two retention mechanisms above T_g the shape of the retention diagram is dependent on experimental conditions. When both surface adsorption and bulk sorption are operative in a column, equilibrium retention volumes are satisfactorily represented for low molecular weight liquids and polymers by the relation:

$$V = K_s W_L + K_a A_L$$

where V is the total retention volume, K_s and K_a the partition coefficients for bulk sorption and surface adsorption respectively, W_L and A_L the mass and surface area of the stationary phase. As the coating thickness (W_L/A_L) is decreased, reversal from the linear behaviour becomes less pronounced, and at low enough coverage a linear retention diagram through T_g is obtained, corresponding exclusively to surface adsorption. The temperature of first deviation from linearity T_1, representing the onset of bulk sorption, was found in excellent agreement with the glass transition temperature of the stationary phase, independent of experimental conditions and thickness of the stationary phase. This is consistent with the concept of the glass transition as an iso-free volume state. The penetration of the problem molecule into the bulk of the polymer provides a sensitive test for the rapid development of excess free volume which is known to occur at or just above the glass transition temperature. Retention volumes were measured at intervals of approximately 3°C with n-pentane on a sample of isotactic polypropylene coated onto an inert support. In view of the temperature range of interest, –40 to + 50°C, a low boiling probe was necessary to produce suitable retention times.

As was the case with amorphous and rubbery polymers, a large and characteristic change in retention behavior was recorded for this highly crystalline (62%) polymer in the glass transition region.

BRITTLE TEMPERATURE

The forced elasticity not only depends on temperature, but also on the rate of stress applied. The temperature where the stress dr becomes zero, rubber-like deformation develops.

At a certain low temperature, the stress required to regroup the chain sections correspond to so short a time that σ_f reaches the brittle strength values σ_{br}. The temperature below which the polymer breaks under the stress σ_{br} is known as brittle temperature.

The brittle temperature can be obtained by plotting curves between σ_f vs T and σ_{br} vs T. The intersection of these two curves will give brittle temperature. Macramolecular glass even on being subjected to the same rate of deformation have different T_g and T_{br}. The difference $T_g - T_{br}$ determines the temperature range of forced elasticity.

EFFECT OF INTERMOLECULAR ATTRACTION ENERGY

In polar glass, macromolecules form strong bonds by crosslinking between the polar groups of the chains which increases its brittle strength. These bonds are also labile and large stresses cause a regrouping of chain sections to load the entire molecular network more uniformly. Thus, the forced elastic deformation occurs at lower temperatures.

EFFECT OF PACKING IN MACROMOLECULES

The loose packing of some of the macromolecules favours the forced elasticity to appear on application of large stresses. On cooling, σ_f does not change appreciably and the brittle temperature is low, *e.g.*, the loose packing of macromolecules of cellulose nitrate and cellulose acetate causes a wide range of forced elasticity.

THE LIQUIDS STATE

In the light of the kinetic molecular theory, liquids may be regarded as a continuation of gases into the region of small volumes and high intermolecular attraction. The cohesive forces in a liquid have been stronger than those in gases even at high pressures, and are sufficiently high to keep the molecules confined to a definite volume. The positions of the molecules in liquids arc not rigidly fixed. These forces are not strong enough to entirely eliminate the movements of the molecules in the liquid.

Properties of liquids can be explained on the basis of their following characteristics :

(a) The forces amongst the constituents of a liquid disallow them from separating spontaneously from each other, but they are not strong enough to hold the molecules in fixed positions. It is possible to explain surface tension, viscosity, fairly high heat of vaporization and vapour pressure of liquids in terms of these attractive forces.

(b) The molecules in a liquid are in a state of random motion although the extent of randomness is appreciably much smaller in comparison to gases. The molecules are relatively close together. Most of the space in the liquid gets occupied by its molecules and only a small fraction of the space is available to them for their free movement. This explains the higher density, incompressibility and slow diffusion of liquids in comparison to gases.

(c) The average kinetic energy of the molecules in a liquid is proportional to the absolute temperature. Increase in temperature would increase the proportion of the energized molecules, lowers the attractive forces between the molecules, and consequently increases the vapour pressure of the liquid.

From me above arguments, it is evident that most of the characteristic properties of the liquid arise due to the nature and magnitude of the intermolecular forces between the molecules. The important properties dealt with in this chapter are :

(i) Vapour pressure,

(ii) Surface tension, and

(iii) Viscosity.

LIQUID-VAPOUR EQUILIBRIUM-VAPOUR PRESSURE

Molecules of a liquid, like those of a gas, exhibit a distribution of energies. The molecules that are having sufficient energy to overcome intermolecular attractions are able to escape from the bulk into the gas phase (or vapour phase) provided they are close to the surface. This happens during evaporation. But the average kinetic energy of molecules in vapour state is more than that in liquid state. Therefore the temperature of the liquid falls on evaporation. The rate of evaporation of a liquid depends upon :

(a) The temperature of the liquid,

(b) Attractive forces in the liquid,

(c) Surface area, and

(d) Pressure above the liquid.

If we examine the evaporation of a liquid in an enclosed space (at constant temperature), it is found that some of the molecules of the liquid will spontaneously pass from the surface into the space above it (Fig. 1.1). But the space is closed. Therefore, the molecules in the vapour state are unable to escape from the container. The molecules in the vapour phase collide with each other and with the sides of the container. Some of the molecules on collision pass on their energy to other molecules and come back to the liquid phase. This phenomenon is termed as condensation. The rate of condensation of molecules in the vapour phase is proportional to the concentration of molecules in the vapour phase. A stage is ultimately reached when the rate of condensation becomes equal to the rate of evaporation. This situation corresponds to that of a dynamic equilibrium between the liquid and its vapour, *e.g.*,

$$H_2O(l) \underset{\text{condensation}}{\overset{\text{evaporation}}{\rightleftharpoons}} H_2O(g)$$

Thus a state of equilibrium may be defined as one in which two opposing processes occur simultaneously at the same rate. At this point there is no further change in the number of molecules in either the liquid or in the vapour state.

Molecules of a liquid in the vapour state exert some pressure. At equilibrium, this pressure is characteristic of the liquid at a given temperature and is called the vapour pressure of the liquid. The vapour pressure of a liquid is therefore defined as the pressure exerted by the vapours that are in equilibrium with the liquid at a given temperature. The vapour pressure has been found to depend on

(i) The nature of the intermolecular forces in the liquid, and

(ii) The temperature of the liquid.

(i) *Nature of the Liquid :* If in a liquid the attractive forces between the molecules are strong then the tendency of the molecules to escape from the surface of the liquid would be low. Hence, such a liquid will have low vapour pressure. However, liquids with low intermolecular forces would be having a high escaping tendency and consequently high equilibrium vapour pressure.

(ii) *Effect of Temperature :* As the temperature of a liquid gets increased the average kinetic energy of the molecules increases. The increased kinetic energy partly overcomes the attractive forces between the molecules and thus raises the escaping tendency of the molecules. This causes an increase in the equilibrium vapour pressure. The values of vapour pressure when plotted against temperature, a curve of the type shown in Fig. 2 is obtained. The curve reveals that the vapour pressure increases exponentially with the increase of temperature. The temperature at which the vapour pressure of a liquid becomes equal to the atmospheric pressure is termed as the boiling point of the liquid. *The boiling point of a liquid is the temperature at which its vapour pressure is equal to the external pressure.* When the external pressure is 1 atmosphere, the term normal boiling point is used.

EQUATION OF STATE FOR LIQUIDS

In gases the distance between the molecules is large as compared to that in solids or liquids. All gases behave nearly alike at very low pressures (strictly speaking $P \rightarrow 0$) so that their behaviour can be adequately explained by an equation of state, $PV = nRT$. Solids and liquids are considered as *condensed phases*. Because of relatively small distances between the molecules, the effects of intermolecular forces are quite predominant in solids and liquids. The magnitude of these forces tend to vary with me nature of the liquid or the solid. Thus, a general equation of state for condensed phases would not be possible.

An equation of state for solids and liquids can be deduced in terms of coefficients of thermal expansion (a) and compressibility (k). These are defined as

$$\alpha = \frac{1}{V}\left(\frac{\partial V}{\partial T}\right)_P \qquad ...(1)$$

$$k = -\frac{1}{V}\left(\frac{\partial V}{\partial T}\right)_T \qquad ...(2)$$

where α refers to the isobaric relative increase $\left(\frac{\partial V}{V}\right)$ in volume per degree rise in temperature while k isothermal relative decrease $\left(-\frac{dV}{V}\right)$ per unit rise in pressure, a and k depend on the nature of the condensed phases and are different for different substances.

For small temperature intervals equation (1) can bc written as follows :

$$\frac{dV}{V} = \alpha\, dT$$

Integrating the above equation within limits, we get

$$\int_{V_0}^{V_t} \frac{dV}{V} = \int_0^T dT$$

$$\text{In} \frac{V_t}{V_0} = \alpha t$$

or $\qquad V_t = V_0\, e^{\alpha t}$

where V_t and V_0 refer to the volumes at t° and 0°C respectively. Expanding $e^{\alpha t}$ in the power series, we get

$$e^{\alpha t} = 1 + \alpha t \left(\frac{(\alpha t)^2}{2!}\right) + ...$$

As α is very small, square and other higher order terms may be neglected. Therefore, we can write

$$V_t = V_0(1 + \alpha t) \qquad ...(3)$$

Equation (3) relates the volume of a substance with temperature, α is constant over a limited temperature range and is approximately constant for gases.

In case of liquids or solids, α is having its own characteristic value, α is having a positive value for gases and solids while for most of the liquids it is positive with some exception, *e.g.*, water between 0 – 4°C has negative value of α.

Similarly, integrating equation (2) we obtain

$$\int_{V_0^0}^{V^0} \frac{dV}{V} = -\int_1^P k dP$$

$$\ln \frac{V_0}{V_0^0} = -k(P-1)$$

or $$V_0 = V_0^0 \, e^{-k(P-1)}$$

where V_0^0 refers to the volume at 1 atmosphere and 0°C.

On expanding $e^{-k(P-1)}$ in the power series and neglecting square and other higher terms, we obtain

$$V^0 = V_0^0 \, [1 - k(P - 1)]$$

Like α, k is always positive and is the same for all gases but each liquid or solid has its own characteristic value of k. k is constant over a fairly large range of pressure. Equation (4) predicts that the volume of a condensed phase gets decreased linearly with increase of pressure.

This behaviour is contrary to the behaviour of gases in which volume is inversely proportional to pressure. Combining equations (3) and (4) and eliminating V_0, we obtain

$$V_t = V_0^0 \, (1 + \alpha t) \, [1 - k(P - 1) \qquad ...(5)$$

The constants a and k in equation (5) depend upon the nature of the condensed phase. Table 2.2 gives the values of α and k for some common liquids and solids.

Table 2.2 : Coefficient of Thermal Expansion and Compressibility at 293 K.

Solid	*Coefficients*		*Liquid*	*Coefficients*	
	$\alpha \times 10^4$ K^{-1}	$k \times 10^6$ atm^{-1}		$\alpha \times 10^{-4}$ K^{-1}	$k \times 10^6$ a^{-1}
Silver	0.582	1.00	C_6H_6	12.4	94.0
Copper	0.491	0.78	CH_3OH	12.0	120.0
Graphite	0.242	3.00	C_2H_5OH	11.2	110.1
Platinum	0.263	0.38	CCl_4	12.4	103.2
NaCl	1.212	4.20	H_2O	2.0	45.3

STRUCTURE OF LIQUIDS

In a perfect crystal a complete ordered arrangement of the atoms, ions or molecules constituting the crystal exists. The intermolecular forces are strong enough to hold them together. The constituents vibrate about their mean positions but cannot execute Translational motion. Hence both the short range as well as the long range orders exist in crystals. The molecules in a gas have complete random motion and the intermolecular forces between the molecules are small and are effective only at short distances. There is no possibility of any kind of ordered structure in gases. In liquids, on the other hand, the situation somewhat lies in between the two. The cohesive forces in liquids are stronger than those in gases. These forces, however, are not strong enough to disallow considerably the translational motion of the individual molecules. In terms of the arrangement of the constituents a liquid is having *short range order* but lacks long range order.

When a crystal melts there is, in general, an increase in volume of about 10% or about 3% in intermolecular spacings. The molecules in the liquid state still remain in the vicinity of the other molecules surrounding them. The intermolecular forces tend to decrease as the distance between the molecules increases. In other words, the short range forces exist while the long range forces are negligible. The thermal motion tends to introduce a disorder in the structure but the motion is not as random as in a gas. Some ordered arrangement still persists. The ordered arrangement in the crystal is not completely destroyed on melting. As the temperature gets further increased, the thermal motions of the molecules increase the kinetic energy, decrease the order until at the boiling point it completely vanishes.

The sharpness of the melting point could be explained by a two dimensional model for solids, liquids and gases as shown in Fig. 6. Introduction of a small region of disorder into a crystal brings about the disturbance of the long range order and destroys the crystalline arrangement. This explains the abrupt variation in the properties between solids and liquids. J.D. Bemal postulated that an atom is surrounded by five atoms (Fig. 1.6) instead of normal six atoms as is the requirement for a closest packed structure for solids. The remaining atoms (circles in the figure) are drawn in the utmost possible ordered arrangement. One point of abnormal coordination number is sufficient to bring about a long range disorder. That is, when the normal motion causes a disorder in one region, it spreads in all directions destroying the entire regular structure. Liquids, therefore, may be regarded to have structure like that of a crystal with a difference that the ordered arrangement extends over a short region instead of over the whole mass. This is termed as the short range order or long range disorder. It is to be noted that the short range order in a liquid structure is continuously changing because of the thermal motions.

X-ray Diffraction of Liquids : It becomes possible to calculate the distribution of atoms or molecules in a liquid by the analysis of the intensity of the scattered X-rays. The diffraction pattern of a liquid resembles a powder photograph with a difference that in place of sharp lines a few broad bands are observed. This reveals the existence of a regular order. From the intensities of these bands, it is possible to construct the radial distribution function which is interpreted in terms of the average number of atoms around a central atom at a certain distance. As the temperature of the liquid gets raised, the bands become less pronounced and the pattern resembles those of a gas revealing a more disordered arrangement at higher temperatures.

Vacancy Model for a Liquid : This model for the structure of liquids was given by Eyring (1933) who assumed the liquids more or less similar to a gas. In the liquid, most of the space gets occupied by the molecules and only a small fraction (about 3% of the total volume at ordinary temperature and pressure) of the total volume is free or void. This void space in which the molecules can move is known as the *free volume.* This is shown in Fig. 1.7. The vapours have only a few molecules moving randomly and thus has large free volume. With a rise in temperature, the concentration of the molecules in the vapour phase increases bringing about an increase in the vacancies of the liquid. The density of the vapour increases while that of the liquid decreases until they become equal at the critical temperature. At the critical temperature the liquid and vapour become indistinguishable. Eyring further regarded that the free volume gets distributed randomly and the vacancies are of approximately molecular size. Molecules adjacent to a vacancy would be expected to posses gas-like properties and show maximum disorder. But a molecule away from the vacancy would possess solid-like properties and possessing a more ordered arrangement. Based on this model, they were able to show a fairly good agreement between the observed and predicted values of the properties.

In a perfect crystal, g is a periodic array of sharp spikes, representing the certainty (in the absence of defects and thermal motion) that particles lie at definite locations. This regularity continues out to the edges of the crystal, so we say that crystals have long-range order. When the crystal melts, the long-range order is lost and, wherever we look at long distances from a given particle, there is equal probability of finding a second particle. Close to the first particle, though, the nearest neighbours might still adopt approximately their original relative positions and, even if they are displaced by new-comers, the new particles might adopt their vacated positions. It is still possible to detect a sphere of nearest neighbours at a distance r_1 and perhaps beyond them a sphere of next-nearest neighbours at r_2. The existence of this short-range order means that the radial distribution function can be expected to oscillate at short distances, with a peak at r_1 a smaller peak at r_2 and perhaps some more structure beyond that the shape of the radial distribution function can be determined by X-ray diffraction, for g can be extracted from the diffuse diffraction pattern characteristic of liquid samples in much the same way as a crystal structure is obtained from X-ray diffraction of crystals.

Closer analysis shows that any given H_2O molecule is surrounded by other molecules at the comers of a tetrahedron, similar to the arrangement in ice. The form of g at 100°C shows that the intermolecular forces (in this case, largely hydrogen bonds) are strong enough to affect the local structure right up to the boiling point.

Calculation of g

Because the radial distribution function can be calculated by making assumptions about the intermolecular forces, it can be used to test theories of liquid structure. However, even a fluid of hard spheres without attractive interactions (a collection of ball-bearings in a container) gives a function that oscillates near the origin and one of the factors influencing, and sometimes dominating, the structure of a liquid is the geometrical problem of stacking together reasonably hard spheres. Indeed, a liquid of hard spheres shows a more pronounced oscillatory behaviour at a given temperature than any other type of liquid. The attractive part of the potential modifies this basic structure but sometimes only quite weakly. One of the reasons behind the difficulty of describing liquids theoretically is the importance of both the attractive and repulsive (hard core) components of the potential. There are several ways of building the intermolecular potential into the calculation of g. Numerical methods take a box of 10^3 particles (the number increases as computers grow more powerful), and the rest of the liquid is simulated by surrounding the box with replications of the original box. Then, whenever a particle leaves the box through one of its faces, its image arrives through the opposite face so that the total number remains constant.

Monte Carlo Methods

In the Monte Carlo method, the particles in the box are moved (usually one at a time) through small but otherwise random distances, and the total potential energy is calculated using one of the intermolecular potentials. Whether or not this new configuration is accepted is then judged from the following rules :

(1) If the potential energy is not greater than before the change, then the configuration is accepted.

(2) If the potential energy is greater than before the change, then it is accepted or rejected with the probability of acceptance in proportion to the value of $e^{-\Delta V_N/kT}$, where ΔV_N is the change in total potential energy of the N particles in the box.

It can be shown that this procedure ensures that at equilibrium the probability of occurrence of any configuration is proportional to the Boltzmann factor $e^{-\Delta V_N/kT}$. The configurations generated in this way can then be analyzed for die value of g simply by counting the number of pairs of particles with a separation r and averaging the result over the whole collection of configurations.

SURFACE TENSION

A molecule in the interior of a liquid is completely surrounded by other molecules and hence, on an average, it is attracted equally in all directions.

However, a molecule on the surface of a liquid is subjected to a resultant inward pull because there is a greater number of molecules per unit volume in the liquid than in the vapour. Because of this inward pull, the surface of a liquid tends to contract to the smallest possible area and behaves as if it were in a state of tension.

Since the natural tendency of a liquid is to decrease its surface area, any increase in the latter can only be accomplished at the expense of work. The side RS, of length *l* is movable. If a force 'f' is required to move RS against the force of surface tension acting in the film along RS, then the work, W, in moving the wire from RS to AB is given by

$$W = f.x \qquad ...(1)$$

This force f will have to be balanced by the force of surface tension along RS. If γ is the force acting per cm along RS due to the surface tension and since there are two surfaces to the film, the force due to surface tension would be $2\gamma l$, then

Work done in increasing the surface area $= 2\gamma.l.x$ ...(2)

or $\quad W = 2\gamma.l..x = f.x$ ...(3)

Hence, $\quad \gamma = \dfrac{f}{2l}$ ...(4)

Therefore, surface tension may be defined as *the force in dyne acting along the surface of a liquid at right angles to any line one cm dn length.*

From equation (3), we have

$$w = \gamma(2lx) = \gamma.\Delta A$$

where ΔA is the increase in surface area of the film. It follows that

$$\gamma = \frac{W}{\Delta A} \qquad ...(5)$$

Therefore, surface tension is the work in ergs required to generate 1 sq. cm of surface area

METHODS OF MEASURING SURFACE TENSION

The methods more frequently used in the determination of surface tension of liquids are described below.

Single Capillary Rise Method

Consider a fine capillary tube of a uniform radius 'r' immersed in a vessel containing a liquid which wets glass. The liquid will rise in the capillary till the force of surface tension f_1 acting upward is equal to the force f_2 due to the column of liquid acting downward.

Let 'γ' be the surface tension of the liquid expressed in dynes per cm of the inner circumference. This force is effective in pulling the liquid upward. Since this force is acting around the circumference, the force acting upward will be given by

$$f_1 = 2\pi r.\gamma \cos\theta \quad ...(6)$$

This upward force is balanced by the weight of the liquid column. Weight of the liquid column,

$$f_2 = \pi r^2 h\rho g \quad ...(7)$$

where 'ρ' is the density of the liquid and 'g' is the Gaseous State due to gravity.

At equilibrium both the forces become equal and the height of the liquid remains steady.

Hence, $2\pi r.\gamma \cos\theta = \pi^2.h.\rho.g$

$$\therefore \quad \gamma = \frac{r h \rho g}{2 \cos\theta} \quad ...(8)$$

The angle θ is often called an angle of contact. For most liquids which wet glass, θ is nearly equal to zero.

$$\therefore \quad \cos\theta = 1$$

$$\theta \quad \gamma = \frac{r h \rho g}{2} \quad ...(9)$$

For liquids which do not wet glass, *e.g.*, mercury, the angle of contact θ has to be taken into consideration.

In order to calculate γ by means of equation (8) or (9) r has to be determined by a travelling microscope, A by means of a cathetometer and ρ by a specific gravity bottle.

Double Capillary Rise Method

This method is useful when the given liquid is available in a small quantity.

If two capillary tubes of radii r_1 and r_2 ($r_2 > r_1$), are immersed in the same liquid whose surface tension is to be determined then the liquid rises to different heights in the tubes. Let the height to which the liquid rises in the two capallaries be h_1 and h_2 cm respectively.

Assuming that θ, the angle of contact, is negligible, it follows from equation (9)

$$\text{Surface tension} = \gamma = \frac{r_1 h_1 \rho g}{2}$$

Similarly, for second capillary tube with radius r_2

$$\gamma = \frac{r_2 h_2 \rho g}{2}$$

$$\therefore \quad \frac{\gamma}{r_1} = \frac{1}{2} h_1 \rho g$$

and $$\frac{\gamma}{r_2} = \frac{1}{2} h_2 \rho g$$

$$\therefore \quad \gamma \left[\frac{1}{r_1} - \frac{1}{r_2}\right] = \frac{1}{2}(h_1 - h_2)\,\rho g$$

Hence if r_1, r_2 and the difference between two heights are known, γ can be calculated from equation (10).

Drop Weight Method

The principle of this method consists in determining the weight of a single drop of a liquid which falls under its own weight through a capillary tube of uniform radius. At the instant the drop is about to fall the force f_1 due to surface tension pulling the drop upwards will be $2\pi r\gamma \cos\theta$. This force is counterbalanced by the weight of the drop f_2 acting downwards. At equilibrium both these forces become equal and hence $2\pi r\,\gamma \cos\theta = mg$, where 'm' is the mass of the drop.

Consider two different liquids having surface tensions γ_1 and γ_2 flowing through the same capillary tube. On assuming that their angles of contacts are equal, we get

$$\frac{2\pi r \gamma_1 \cos\theta}{2\pi r \gamma_2 \cos\theta} = \frac{m_1 g}{m_2 g}$$

$$\therefore \quad \frac{\gamma_1}{\gamma_2} = \frac{m_1}{m_2} \qquad \text{...(11)}$$

Hence, if we determine the masses of a single drop of each of the liquids, and knowing the surface tension of any one liquid that of the other can be calculated.

Drop Number Method

Since it is very difficult to determine the weight of a single drop, the drop weight method is not very convenient. In the drop number method, the number of drops formed from a fixed volume of a liquid (between marks A and B in Fig. 1.15) is determined by using an instrument called the Traube Stalagmometer. The lower portion of the stalagmometer consists of a fine capillary tube through which the liquids can flow. The upper portion is connected by rubber tubings to a Y-shaped glass tube.

At the start, the screw clamp is closed and the given liquid is sucked into the stalagmometer to a level above mark A by opening the pinch cock. The pinch cock is then closed and the screw clamp gradually opened, so that number of drops flowing out is between 15 and 20 per minute.

Once this drop rate is achieved the screw clamp is left undisturbed. The liquid is then again sucked in by opening the pinch cock without allowing any air bubble to remain in the capillary. The number of drops formed while the liquid flows from the mark A to B is then determined. Let this number be denoted by n_1.

If $2\pi r\gamma_1 \cos\theta$ is the force due to surface tension on a single drop, then the force on n, drops will be $2\pi r\gamma_1 n_1 \cos\theta$. This latter force will have to be balanced by the weight of n_1 drops which is M_1g where M_1 is the mass of n_1 drops.

$$\therefore \quad 2\pi r n_1\gamma_1 \cos\theta = M_1 g$$

$$\therefore \quad 2\pi r n_1\gamma_1 \cos\theta = V.\rho_1.g \qquad ...(12)$$

where V = volume of the liquid between the marks A and B,

ρ_1 = density of the liquid.

The stalagmometer is then cleaned and dried and the same procedure repeated with another liquid of surface tension γ_2. Let the number of drops obtained with this liquid be denoted by n_2.

$$\therefore \quad 2\pi r\, n_2\gamma_2 \cos\theta = M_2 g$$

$$\therefore \quad 2\pi r\, n_2\gamma_2 \cos\theta = V.\rho_2.g \qquad ...(13)$$

From equations (12) and (13) we have,

$$\frac{2\pi r_1\, n_1\gamma_1 \cos\theta}{2\pi r_2\, n_2\gamma_2 \cos\theta} = \frac{v.\rho_1.g}{v.\rho_2.g}$$

$$\therefore \quad \frac{n_1}{n_2} = \frac{\rho_1}{\rho_2} \times \frac{\gamma_2}{\gamma_1}$$

$$\therefore \quad \gamma_2 = \frac{n_1}{n_2} \cdot \frac{\rho_2}{\rho_1} \cdot \gamma_1 \qquad ...(14)$$

Hence, if n_1 and n_2 are determined and so also the densities of the two liquids, then the surface tension of any one liquid can be calculated if that of the other liquid is known.

VISCOSITY

Liquids possess another property known as viscosity which implies resistance to flow. It is developed in liquids because of the shearing effect of moving one layer of liquid past another.

Consider a liquid to be divided into parallel planes or layers of molecules, a fixed distance apart. Let the area of each plane be A and the distance between planes be dy. Suppose each plane is moving to the right with velocities v_0, v_1, v_2 etc. such that each succeeding velocity is greater than the preceding by an amount dv. This type of flow is called *laminar flow* or *streamlined flow as distinct from turbulent or vortex flow* in which parallelism of the layers is not maintained.

In laminar flow the force 'f' required to maintain a steady velocity difference 'dv' between any two parallel planes is directly proportional to the area of plane A and the velocity difference between the two planes dv. It is also inversely proportional to the distance between the planes.

Mathematically, it can be expressed as

$$\therefore \quad f \propto A.\frac{dv}{dy}$$

$$\therefore \quad f = \eta.A.\frac{dv}{dy} \qquad ...(1)$$

where, η is called the coefficient of viscosity of the liquid.

$$\therefore \quad \eta = \frac{f}{A \times \frac{dv}{dy}} \qquad ...(2)$$

When $A = 1$ sq.cm and $dv = 1\text{cm sec}^{-1}$,

$dy = 1\text{cm}$ then η = force in dynes

$$\text{units } \eta = \frac{\text{dynes}}{\text{cm}^2 \times \frac{\text{cm/sec}}{\text{cm}}} = \text{dynes.sec.cm}^{-1}$$

This is the absolute unit of viscosity coefficient. The practical unit in the C.G.S. system is known as poise.

$\therefore$ 1 poise $\equiv$ 1 dynes . sec cm^{-1}.

Other units in use are centipoise (0.01 *poise*) and *millipoise* (0.001 *poise*).

Therefore, the coefficient of viscosity (η) 'Can be defined' as the force in dynes that must be exerted between two parallel planes 1 cm^2 in area and 1 cm apart in order to maintain a laminar velocity difference of 1 cm sec of one layer past another.

METHODS OF DETERMINING VISCOSITY COEFFICIENT

The method commonly used for determining viscosity coefficient is based on Poiseuilles equation,

$$\eta = \frac{\pi P r^4 t}{8 L V} \qquad ...(3)$$

where V = volume (in cc) of a liquid of viscosity η that flows through a capillary tube of radius r cm and length L cm in time t secs under a pressure head of P dynes/cm^2.

Instead of determining η by a direct application of the Poiseuille equation, the one commonly used is the one devised by Ostwald in which the viscosity of one liquid is compared with that of a liquid of known viscosity. (*e.g.*, H_2O).

Use of Ostwald Viscometer

A known volume (say V_0 cc) of the given liquid is introduced into bulb B of the viscometer. The liquid is then gently sucked into the bulb A above the mark 'x' and the time (t_2) required for the liquid to flow through the capillary from 'x' to 'y' is accurately determined using a stop watch. A mean of three observations is recorded. The viscometer is then cleaned and dried and the same volume (V_0 cc) of water, of known viscosity, is introduced into bulb B. This is again sucked into bulb A gently and the time (t_1) required for water to flow between marks 'x' and 'y' is determined as before.

From equation (3), we have,

For water $\eta_1 = \dfrac{\pi P_1 . r^4 . t_1}{8 L V}$

For the given liquid $\eta_2 = \dfrac{\pi P_2 . r^4 . t_2}{8 L V}$

Note that V = volume corresponding to that between the marks 'x' and 'y'

$$\therefore \quad \frac{\eta_2}{\eta_1} = \frac{P_2.t_2}{P_1.t_1} \quad ...(4)$$

But $$P = \frac{\text{Force}}{\text{Area}} = \frac{mg}{A} = \frac{V_0.\rho g}{A}$$

$$\therefore \quad \frac{P_2}{P_1} = \frac{V_0.\rho_2\, g/A}{V_0.\rho_1\, g/A} = \frac{\rho_2}{\rho_1}$$

From equation (4), we get

$$\frac{\eta_2}{\eta_1} = \frac{\rho_2.t_2}{\rho_1.t_1}$$

$$\therefore \quad \eta_2 = \frac{t_2}{t_1} \times \frac{\rho_2}{\rho_1} \times \eta_1 \quad ...(5)$$

Since all quantities on the R.H.S. of equation (5) are known, η_2 can be calculated. This is called the *absolute viscosity* of the given liquid. Then η_2/η_1 is the relative viscosity of liquid 2 w.r.t. liquid 1. If liquid 1 is water, then instead of using the term "relative viscosity", for the ratio η_2/η_1 we use the term "*specific viscosity*".

Determination of η by the 'Falling Sphere' Method

This method is specially useful for liquids having a high coefficient of viscosity.

Principle : A spherical body of a radius r and density p falling under gravity through a liquid medium of density ρ is acted upon by the gravitational force f_1 which is given by

$$f_1 = \frac{4}{3}\pi r^2(\rho - \rho')g \quad ...(6)$$

This force which tends to accelerate the body falling through the liquid is opposed by the frictional forces within the liquid which increase as the velocity of the falling body increases. Ultimately, a uniform rate of fall is attained at which the frictional forces become equal to the gravitational force and thereafter the body continues to fall with a constant velocity v called the *terminal velocity*. Stokes showed that this frictional force f_2 is given by

$$f_2 = 6\pi r\eta v \quad ...(7)$$

where r = radius of sphere

v = terminal velocity

η = coefficient of viscosity of the liquid.

Equating the two forces, we get

$$\frac{4}{3}\pi r^3(\rho - \rho')g = 6\pi r\eta\upsilon$$

$$\therefore \quad \eta = \frac{2r^2(\rho-\rho')g}{9v} \quad \text{...(8)}$$

$$\frac{\eta_1}{\eta_2} = \frac{(\rho-\rho_1')}{(\rho-\rho_2')} \times \frac{v_2}{v_1}$$

But $$v = \frac{\text{distance between 'x' and 'y'}}{\text{time of fall}}$$

$$\therefore \quad \frac{\eta_1}{\eta_2} = \frac{(\rho-\rho_1')}{(\rho-\rho_2')} \times \frac{t_1}{t_2} \quad \text{...(9)}$$

EFFECT OF TEMPERATURE ON VISCOSITY

It has been found that the viscosity of a liquid decreases markedly with rise in temperature. The variation of viscosity with temperature is best expressed by the equation.

$$\eta = A.e^{E/RT} \quad \text{...(1)}$$

where A and E are constants for a given liquid.

Taking logarithms

$$ln\ \eta = ln\ A + \frac{E}{RT}$$

$$\therefore \log_{10}\eta = \log_{10} A + \frac{0.4343\ E}{RT} \quad \text{...(2)}$$

Hence, a plot of $\log_{10} \eta$ versus $1/T$ should be linear. This has actually been borne out for a large number of liquids.

The reason why viscosity decreases with temperature is that before a molecule can take part in liquid flow it must acquire a sufficient energy to push aside the molecules surrounding it. As the temperature increases the number of surrounding molecules increases in proportion to the Boltzmann factor $e^{-E/RT}$ and hence the resistance to flow *e.g.*, viscosity may be expected to decrease according to the factor as $e^{-E/RT}$ equation (1).

LIQUID—LIQUID EQUILIBRIUM

(a) *Introduction :* Two liquids may be completely miscible, partially or completely immiscible with each other. Water and alcohol form an example of the first type, mercury and water constitute an illustration of the third type. Ether and water belonging to the second of the above categories, when mixed together form to separate layers, one being an aqueous solution of ether and the other, ethereal solution of water.

The two solutions thus formed are known as conjugate solutions. It is with such cases of partial miscibility of liquids that we shall be concerned, because they represent this type of equilibria faithfully. Complete miscibility yields only one phase, while we are going to consider the cases of two components in two phases and so they are left out from our considerations. Complete immiscibility, however, yields two phases no doubt, but each phase is constituted of the pure component (only one) and as such loses interest in this connection.

(b) *Classification of L-L System :* L-L systems can be classified into three different classes:

(i) The liquid pairs completely miscible in all proportions, *e.g.*, water-alcohol, benzene-nitrobenzene etc.

(ii) The liquid pairs partially miscible with one another, *e.g.*, phenol-water, nicotine-water etc.

(iii) The liquid pairs completely immiscible with one another, *e.g.*, water-benzene, alcohol- nitrobenzene etc.

Completely Miscible Liquid Pairs

There are certain pairs of liquids which dissolve in each other in all proportions. The general rule about solubility has been that closer the chemical similarity of the two liquids, the greater will be their miscibility. The expression, *"Similia Similibus Solvunter"* (like dissolved by like) is fully justified.

It is evident that the partial pressure of the components can be determined by the composition of the liquid and its temperature. This statement is present in *Duhem-Margales* equation for two components A and B, according to which

$$\frac{d \log P_A}{d \log x_A} = \frac{d \log P_B}{d \log x_B} \quad ...(1)$$

The terms x and p denote the mole fraction and partial pressure of the respective components.

Derivation of Duhem-Margules Equation

Suppose there is a binary solution which is having n_B moles of A and n_A mole of B. Applying *Gibbs-Duhem Equation*, we get at equilibrium:

$$n_A d\mu_A + n_B d\mu_B = 0 \quad ...(2)$$

were μ_A and μ_B denote the chemical potentials of A and B.

On dividing (2) by $(n_A + n_B)$, we obtain

$$\frac{n_A}{n_A + n_B}.d\mu_A + \frac{n_B}{n_A + n_B}.d\mu_B = 0 \quad ...(3)$$

where, x_A and x_B denote the mole fractions of constituents A and B respectively.

If the liquid mixture of A and B has been in equilibrium with their vapour phase at constant temperature T, the chemical potentials would be the same in both the phases for each component. Since,

$$\mu = RT \log f_i + \mu_i^0 \qquad \text{(where } f_1 = \text{fugacity)}$$

or if the vapours are to behave ideally, we can write

$$\mu_i = RT \log p_i + \mu_i^0$$

where P denotes partial pressure of ith component in the vapour phase.

Thus $\mu = RT\ d \log p_i$.

On substituting it in equation (3), we get

$$x_A\ RT\ d \log p_A + x_B\ RT\ d \log p_B = 0$$

or $$\frac{x_A\, d \log p_A}{d x_A} + \frac{x_B\, d \log p_B}{d x_B} = 0$$

$\because$ $x_A + x_B = 1,\ dx_A = -dx_B.$

$\therefore$ $$\frac{x_A\, d \log p_A}{d x_A} + \frac{x_B\, d \log p_B}{d x_B} = 0$$

$$\frac{x_A\, d \log p_A}{d x_A} + \frac{x_B\, d \log p_B}{d x_B}$$

or
$$\frac{d\log p_A}{d\log x_A} = \frac{d\log p_B}{d\log x_B} \quad ...(4)$$

Equation (4) is known as *Duhem-Margules Equation.* It gives rise to a correlation between the composition in the liquid phase and the partial vapour pressures in the gaseous phase.

Konowaloffs Rule

It is also possible to write the *Duhem-Margules Equation* (4) in the form

$$\frac{x_A}{p_A}.\frac{dp_A}{dx_A} = \frac{x_B}{p_B}.\frac{dp_B}{dx_B} \quad ...(5)$$

or since $dx_A = -dx_B$

$$\left[x_A = \frac{n_A}{n_A + n_B},\ x_B = \frac{n_B}{n_A + n_B},\right.$$

where n denotes number of moles of the components

$$x_A + x_B = \frac{n_A}{n_A + n_B} + \frac{n_B}{n_A + n_B} = 1$$

or $\quad x_A = -x_B$

or $\quad dx_A = -dx_B]$

Hence, (4) becomes as follows:

$$\frac{x_A}{p_A}.\frac{dp_A}{dx_A} + \frac{x_B}{p_B}.\frac{dp_B}{dx_B} = 0$$

Hence.
$$\frac{dP}{dx_A} = \frac{dp_A}{dx_A} + \frac{dp_B}{dx_A}\left(1 - \frac{x_B \cdot p_A}{x_A \cdot p_B}\right), \quad ...(6)$$

where P = total pressure.

The factor $\frac{dp_B}{dx_A}$ must be negative, because it is equal to $\frac{-dp_B}{dx_B}$.

If $\frac{dp}{dx_A}$ is positive, then we have

$$x_B\, p_A > x_A\, p_B$$

or
$$\frac{p_A}{p_B} > \frac{x_A}{x_B} \quad ...(7)$$

If $\frac{dp}{dx_A}$ is negative *i.e.* $\frac{dp}{dx_B}$ is positive, then we have

$$x_B P_A < x_A p_B$$

or $$\frac{p_A}{p_B} < \frac{x_A}{x_B} \quad ...(8)$$

Inequalities (7) and (8) have been the mathematical expressions of *Konowaloff's rule* that the vapour is richer in the component whose addition to the liquid mixture causes an increase of total vapour pressures.

Ideal Solutions of Liquids

An ideal solution may be defined as one in which the partial pressure of each component is proportional to the molar concentration in the mixture. This is also called *Raoult's law*. It may be mathematically stated in the form

$$p_A = p_A^0 \cdot x_A,$$
$$p_B = p_B^0 \cdot x_B,$$

where,

p_A, p_B = partial pressure of the respective components A and B

p_A^0, p_B^0 = vapour pressure of the two components

x_A, x_B = molar fraction of the two components.

Thus this law may be stated that the partial vapour pressure of any volatile constituent of a solution would be equal to its vapour pressure in the pure state multiplied by the mole fraction of two constituents in the solution.

DETERMINATION OF RATIO OF DISTILLATION TO RESIDUE

The temperature-composition distillation diagram makes us to calculate the proportion of weight of distillate to weight of residue at a given temperature. Suppose we start with a liquid mixture of weight w having a composition of x% B. The liquid will start boiling only when the temperature gets raised to T_1°. With vaporisation, the liquid will get more concentrated in A and the boiling temperature will rise. At a temperature, say T_2°, the liquid will possess the composition as at y_1 which will be having x_1 % B and suppose the weight of the liquid now is w_1. If the system is closed, the vapour phase will be having the composition as at y_2, having x_2% B; the weight of the vapour phase is w_2.

Thus, $w = w_1 + w_2$, and the distribution of B would be such that

$$wx = w_1x_1 + w_2x_2$$

or $$(w_1 + w_2)\,x = w_1x_1 + w_2x_2$$

or $$w_1\,(x - x_1) = w_2\,(x_2 - x)$$

or $$\frac{w_1}{w_2} = \frac{x_2 - x}{x - x_1}$$

From the diagram, we can see

$$\frac{x_2 - x}{w - x_1} = \frac{yy_2}{yy_1}$$

Hence, $$\frac{\text{weight of distillate}}{\text{weight of residue}} = \frac{w_2}{w_1} = \frac{yy_1}{yy_2}$$

= ratio of the intercepts of the tie line y_1y_2.

Azeotropic Mixtures

A mixture which is possessing a lower vapour pressure (higher boiling point) or higher vapour pressure (lower boiling point) than any other mixture is termed as an azeotropic mixture. Such mixtures separate out on distillation into one of the components and a constant boiling mixture of both. Azeotropic mixtures are also termed as constant boiling mixtures possessing definite composition under a constant pressure.

If the vapour pressure-composition curve is having a minimum or maximum, then from equation (v), we have

$$\frac{dP}{dx} = 0.$$

Hence, $$x_Ap_A = x_Bp_B$$

or $$\frac{x_A}{x_B} = \frac{p_A}{p_B}$$

The other alternative $\frac{dp_B}{dx_A} = 0$ would be unlikely, because it would imply that the partial pressure would happen to remain constant in spite of a change in concentration. As the molecular ratio of the two components in the vapour may be given by $\frac{p_A}{p_B}$, if ideal behaviour is regarded, the composition of the vapour would be the same as that of the liquid with which it remains in equilibrium.

On applying phase rule in azeotropic mixtures, we should remember that there exists only one restriction viz; compositions in liquid and in vapour phase are the same.

Hence, $F' = C - P + 1 = 2 - 2 + 1 = 1$

i.e., azeotropes should behave as univariant systems, so that the boiling point will remain constant, if pressure is fixed. This is what is seen in the next two cases.

(a) Distillation of Liquid Pairs with Maximum Boiling Point

The boiling temperature composition diagram is depicted in Fig. 26. The constant boiling mixture is having the maximum boiling point, *i.e.*, it is the least volatile. The vapour phase for any mixture lying between A and M would, therefore, get richer in A and mixture lying between B and M, would be richer in B than is the constant boiling mixture M. Suppose, a mixture of composition x is make to distil. The first fraction distilled will be having composition indicated by x_1. Evidently, it is richer in A. The composition of the residual liquid, thus shifted towards constant boiling mixture M. As the distillation gets continued, the composition of the distillate gets changed towards A and that of the residue towards M. Ultimately, a distillate of pure A and a residue of constant boiling mixture M would be obtained.

Similarly, a mixture of composition lying between B and M, say y on distillation will ultimately provide a distillate of pure B and a residue of constant boiling mixture M.

It, therefore, follows that any binary solution of this type, on complete fractional distillation, can get separated into a residue of composition M and a distillate of either A and B depending upon whether the initial composition is lying between A and M at between B and M respectively. Thus, it becomes not possible to completely separate such a binary mixture into pure components A and B on distillation.

This mixture having the maximum boiling point is termed 'as *Maximum Boiling Azeotrope* and behaves as if it is a pure chemical compound of two components, as it boils at a constant temperature and the composition of the liquid and vapour is the same. But the azeotrope has been not a chemical compound, as its composition is not constant under all conditions and rarely corresponds to stoichiometric proportions.

Pure water and hydrogen chloride are made to boil at 100° and –85° while their constant boiling mixture (azeotropic mixture) having 20.25% of hydrogen chloride boils at 108.5°, under a pressure of 1 atmosphere. If a solution having less than 20.25% of HCl is made to distil (*i.e.*, between points A and M), water will pass over as the distillate and the residue left behind in the flask would be having 20.25% solution of HCl in water. Thus pure HCl cannot be obtained. Similarly, if a solution having more than 20.25% HCl is distilled, then pure HCl will pass over as distillate and the residue left behind in the flask will be having a mixture of the same constant composition, viz., 20.25% HCl in water.

(b) Distillation of Liquid Pairs with Minimum Boiling Point

If the distillation of composition represented by x is done then the first fraction collected will be having the composition x_1. It will be richer in the constant boiling mixture. The composition of the residual liquid will get shifted towards A. As distillation continues the composition of the distillate and residual liquid gets changed towards M and A respectively. By repeating this process, we get the mixture of minimum boiling point of composition M as distillate, while the residue left over in the distillation flask will be having only pure liquid A.

If we distil a liquid of composition represented by y, then the composition of the first fraction would be represented b y_1. Evidently, it will get richer in the constant boiling mixture. The composition of the liquid will gets richer in B. As the distillation carries out, the distillate and the residual liquid will get richer and richer in constant boiling mixture and pure B respectively. Finally, the distillate will be having only the constant boiling mixture and the residual liquid in the distillation flask will be having only B, There will occur no pure A in this case. *If the mixture is having the azeotropic composition (say z), it will distil unchanged.* In the system of water-ethanol the point M would correspond to a minimum boiling temperature of 78.13 and a composition of 95.57% ethanol by weight. If any solution of composition between pure water and 95.57% ethanol is made to distil, then we obtain a residue of pure water and a constant minimum boiling mixture of 95.57% alcohol in the distillate. No pure ethanol can be recovered. On the contrary, when a solution of composition between pure alcohol and 95.57% ethyl alcohol is made to distil, then we obtain mixture having 95.57% ethanol and pure alcohol. No pure water will yet recovered.

Binary Mixtures

Table 2.3 : (Minimum boiling. point).

Components *A*	*B*	*Min. b. pt.*	*Composition % of B*
H_2O	Pyridine	92.6	59
H_2O	C_2H_5OH	78.13	95.57
C_6H_6	CH_3COOH	80.55	2.0
CS_2	Ethyl acetate	46.0	3.0

Partially Miscible Liquid Pairs

It is regarded that the actual relation between the partial pressures of the components would be given by

$$P_A = P_A^{\circ}.x_A.e^{\alpha}.x^2B$$

$$P_B = P_B^{\circ}.x_B.e^{\alpha}.x^2_A$$

where, a is a constant and all others are having their usual meaning.

If two liquids form a mixture and show a positive deviation from Raoult's law, the value of a is positive. On cooling the system, the positive deviation gets increased and then a also increases. It can be then by differentiating p by taking general case, with respect to x and equating it to zero. On rearranging, we obtained

$$2\alpha x^2 - 2\alpha x + 1 = 0$$

or $$x = \frac{2\alpha \pm \sqrt{4\alpha^2 - 8\alpha)}}{4\alpha}$$

The roots have been real if $(4\alpha^2 - 8\alpha)$ has been positive, *i.e.*,

$$4\alpha^2 - 8\alpha > 0,$$

or $$\alpha - 2 > 2,$$

or $$\alpha > 2.$$

The roots will be equal if $\alpha = 2$.

The roots would be imaginary when $\alpha < 2$. The corresponding curves are depicted in the given

The curve, a, b and c would correspond to increasing temperatures. At the highest temperature, the deviation is appreciably small (curve c),. At the lowest temperature, curve a exhibits a maximum and minimum. The S shaped curve reveals that the three different solutions separated by x, y and z should be having the same vapour pressure.

Hence, only two liquid phase would be in equilibrium with vapour. Hence, three liquid phases cannot exist and in particle the liquid would break up into two liquid layers of composition x and y.

If the temperature is raised, points x and z come closer and closer, till at a certain temperature, the two points merge into one. In that case $\alpha = 2$. This particular temperature is termed as *upper consulate temperature* of upper critical solution temperature (U.C.S.T.). At this temperature, the two liquid phases would be becoming identical. Above this temperature, the liquids will get miscible in all proportions. At the consulate temperature, the system would be non-variant

So, if pressure is maintained constant temperature will be a fixed point.

Partially Miscible Liquid Pairs

There are quite a number of liquid pairs which show only *limited miscibility* in each other *i.e.*, they are only partially miscible. This is similar to the solution of a sparingly soluble salt in a liquid. For example, when a small amount of phenol or aniline is added to water at room temperature and shaken, a solution of phenol or aniline in water is obtained. However, on adding a larger amount of phenol or aniline, two liquid layers separate. One is a solution of phenol or aniline in water and the other a solution of water in phenol or aniline. *Each layer is a saturated solution of one in the other*. At constant temperature, the composition of the two layers, although different from each other, remain constant as long as the two phases are present. The two layers in equilibrium are called *conjugate solutions*. The addition of any one of the liquids merely changes the relative volumes of the two layers and not their composition. Further addition of phenol or aniline causes the water layer to diminish in size which finally disappears and only a layer of water in phenol or aniline remains. The mutual solubility of the two liquids varies in a characteristic manner with temperature.

THE SOLID STATE

Solid, liquids and gas are the three states of matter. Thus, matter exist in three physical stares; gas, liquid and solid. Solids under ordinary conditions have definite volume and shape and tend to maintain these even under deforming forces. Liquids also have definite volume but take the shape of the vessel into while they are poured. Both the liquids and solids are nearly incompressible. Gases, on the otherhand, maintain neither the volume nor shape and completely fill the container into which they are introduced. Gases can be expanded or compressed very easily.

The three states of matter are interconvertible. This can be done by heating or cooling. Heating increases the interparticle spacing and the kinetic energy of the particles. So a solid on interparticle spacing and the kinetic energy of the particles. Many properties of solids, liquids and gases can be easily observed with the help of our sense organs. The properties which can be observed with the help of our sense organs are called macroscopic properties.

The description of the behaviour of the three states of matter an terms of atomic theory is called microscopic description of matter. In chemistry, we try to explain the macroscopic behaviour of matter in terms of its microscopic description. In this chapter, we would try to explain the observable properties of different states of matter in terms of the behaviour of the constituent particles in term.

In a solid, the constituent particles atms, ions or molecules are most closely packed and have the strongest inter molecular force of attraction. As a result particles are fixed in their position and do not have any freedom of motion except that of vibration about their mean position. This give rise to the following characteristic properties to a solid.

(i) A definite shape and definite volume

(ii) Incompressibility

(iii) Rigidity

(iv) No fluidity

(v) Poor diffusibility through them

SOLIDS CHARACTERISTIC

Some characteristic properties of solids are given as follows :

(i) Solids maintain their volume independent of the size of shape of the container in which they are placed.

(ii) Solids are rigid and have definite shape.

(iii) Solids diffuse very slowly compared to liquids or gases. This is because the constituent particles in solids are closely paced and there is very little space to permit the movement of the particles.

(iv) Solids re nearly incompressible. The compressibility of solids is about 10th times that of the gases.

(v) Most solids melt on heating. Some undergo sublimation upon beating. Melting temperatures vary considerably from solids to solid.

(vi) Solids have very high mass-to-volume ratio (density) as compared to gases.

CLASSIFICATION OF SOLIDS

Polycrystalline Solids

In certain crystalline solids, the crystals are very find. These cannot be seen by the naked eyes. Such solids given an impression of being amorphous. Such fine crystalline solids, which appear as amorphous are termed as polycrystalline solids. Metal powders are polycrystalline in nature.

In a polycrystalline sample, there are small crystals. These crystals are randomly oriented. As a result, a sample of polycrystalline material appears to be isotropic, even when each individual crystal is anisotropic.

Amorphous Solids

Amorphous solids have some mechanical properties which are commonly associated with the word solid. But, the amorphous solids differ from the crystalline solids in many respects. So, amorphous solids are sometimes called as pseudo solids.

Some characteristic properties of the amorphous solids are:

(i) The mechanical, electrical and optical properties of amorphous solids do not depend upon the direction, *i.e.*, amorphous solids are isotropic. In this respect, the amorphous solids resemble liquids. It is for this reason that the amorphous solids are considered as supercooled liquids. There is only a short-range order in amorphous solids.

(ii) Amorphous solids do not poses sharp melting points. This indicates the absence of long-range order Amorphous soften on heating and gradually begin to flow like liquids.

(iii) Amorphous solids do not occur in characteristic geometrical shapes.

(iv) When hammered, amorphous material break in an irregular manner. When you hammer a glass sheet or a plastic sheet, these get broken in a very irregular way along many directions.

Examples. Some typical amorphous materials are Plastics, Glass, Rubber, Lamp black, pitch etc.

The solid state is next to gaseous state in its simplicity to understand, although it is also not free from complex problems. Solids are obtained by cooling liquids. There are inherent two possible ways of passing from the equilibrium liquid state into solid state namely crystallization and glass transition.

Crystallization is a transition from a state of short range order to one of long range order in the process of formation of a new phase. The concept of short range order and long range order is defined by the ratio between the distance over which the order extends and the dimensions of the elements arranged in the order.

This is brought about by slowly cooling the liquid where the molecules, due to enough thermal energy and low viscosity of the medium, get enough opportunity to arrange themselves in a geometric order. With the slow release of energy, they grow on the nucleus maintaining the shape of the crystal. It is a phase transition of the first order.

Glass transition does not involve a change of phase, retaining a short range order. It is actually a super-cooled liquid in a non-equilibrium state, which solidify on cooling passing into glassy state. It loses the property of liquid state and acquires the properties of solid state. These changes do not occur abruptly but gradually over a certain range of temperature. The effect of heating an amorphous and crystalline solid. It can be seen that crystals have a sharp melting point while amorphous solids soften over a range of temperature.

Crystalline Solids

All crystalline solids have the following characteristics.

(i) The constituent particles in crystals are generally held by strong interatomic interionic or intermolecular forces. Particles in a solid do not possess translation motion, but they vibrate about their equilibrium positions.

(ii) Crystalline solids possess characteristic geometrical shapes.

(iii) When cut or hammered gently, crystalline solids show a clean cleavage, *i.e.*, they show fracture along a smooth surface.

(iv) Crystalline solids have sharp melting points. This indicate that there exists a long range order in crystalline solids. The long range order in crystalline solids is due to regular arrangement of the constituent particles (atoms, ions or molecules) in three dimensions.

Single Crystals

Single or ideal crystals are formed by the parallel transfer of unit cells along edges at distances equal to periods. Fischer first of all ascertained the possibility of high molecular compounds to form monocrystals. Polymer single crystals like other crystals have been found to contain certain defects originating in the packing of chains.

Lamellar Crystals

Polymer crystals are grown by crystallizing from a one per cent solution of polymer under gradual cooling or isothermal aging under the temperature of dilution. The dimensions, shape and regularity of structure of a single crystal depend on the chemical structure of a chain and the physical constraints like temperature, concentration of solution, the cooling rate, the nature of the solvent, etc. The polymer single crystals are monolayer flat plane Lamella which are often rhomboid of 100Å thick and the size of the plate up 1 mp. The axes a and & of a crystalline cell correspond to long and short diagonal of the rhomboid, while axis c along which the polymer chains are directed is perpendicular to the crystal plane. Since the length of the polymer is very big while thickness of the crystal is of the order of about 100Å, the long chain can only fit in the crystal by turning on its surface by 180°. This is known as chain fold conformation. To have an idea in a crystal of 120Å thickness, the fold shall contain 100 carbon atoms. A molecule having a weight of 105 shall fold about 70 times.

The thickness of a lamella, the dimensions of crystallite and degree of defects of the boundary layers of a crystal are affected by the crystallization conditions. Crystals with folded chain conformation can be obtained when super cooling is in the range 15° – 20°C. Crystallization under high hydrostatic pressure 5-10 × 10^3 atm give extended chain crystals. Polyethylene annealed under a pressure of 700 atm give extended chain crystals. When the melt is cooled under great deformation crystals of different morphological forms can be obtained. Polyhedrons known as hedrites and axialites or avoids are obtained by raising the concentration of the solutions.

Fibrillar Crystals

Fibrillar crystals are obtained by the high rate of evaporation of a solvent from a moderately concentrated solution or at cooling of melt. There are two views about their formation. One section believes that aggregation of rolled lamella is responsible for them while other hold that lamella degenerate in the formation of fibrillar crystals.

In stereospecific polymerization, fibrillar crystals are directly found. The formation is determined by the ratio of the rates of growth of a chain and its crystallization. By changing this ratio crystal either with extended or folded chains are obtained.

Polymer Crystals

Polymers are products of high molecular mass and their decomposition temperatures are far below their boiling points. Hence, they exist only in liquid or solid state. Although, most of the polymers are amorphous, but the crystalline polymers known as isotactic polymers have high melting point and produce fibres of high tensile strength and excellent mechanical properties. A polymer has two types of structural elements viz. monomer units and chains. The long range order is a prerequisite for the crystallization of polymer, not only in the arrangement of units but of chains as well as in three dimensions. This can happen in three ways:

1. Long-range order of arrangement of both chains and monomeric units in three dimensions.
2. Long-range order of arrangement of chains, but the monomeric units are arranged disorderly.

3. Long-range order of the arrangement of units but short range order of chain arrangement.

The high degree of ordering can be achieved in polymers in two ways, *i.e.*, by crystallization or by chain orientation without orientation of monomeric units. These correspond to crystalline and amorphous (glassy) state. It is possible to have a perfect short-range order in chains but the long-range order in arrangement of chains and monomeric units is not very perfect. Therefore, the examples of well ordered liquids and defective crystals are well known.

Crystallization of Polymers

Many polymers are incapable of crystallization under any condition. Their crystallizability is determined by their chemical constitution, *i.e.*, their packing, energy of intermolecular reaction, chain regularity, flexibility, etc.,

Regularity of Polymer Chain

A long-range order must exist in the chain in three dimensions. Sometimes irregularity in the chain may not obstruct crystallization but random and atactic copolymers cannot be crystallized under any circumstances. Natta by using certain catalysts succeeded in synthesizing several highly crystalline vinyl polymers, *e.g.*, polypropylene, polybutylene and polystyrene. A mixture of the polymers of various degree of orientation was formed and crystalline fraction was separated from amorphous fraction by means of solvents as crystalline polymers are less soluble than amorphous polymers.

Second Order Transition Temperature

A first order transition is one at which a definite change in volume occurs. A second order transition is one at which a change occurs only in the rate of change of volume with temperature. Below T_g, polymers are rigid, glass like material and above T_g they are soft elastometric materials. The second order transitions are caused by thermal agitation releasing portions of the resin chain to rotate around its bonds.

These second order transitions can be measured conveniently by deflection temperature DT tests.

Flexibility of Polymer Chain

Polymers with sufficiently flexible chain can crystallize, but liquid chain polymers crystallize with difficulty because the thermal motion of the momeric, units disturb the established order and the crystal breaks. To crystallize a polymer, it is cooled to a temperature where the motion of its units does not hinder their orientation but excessive cooling may also hinder the rearrangement of the units. Crystallization of each polymer is possible only within a definite temperature interval, *i.e.*, between the glass temperature and flow temperature, *e.g.*, the glass temperature of isotactic polystyrene is about 100°C. Hence, it will only crystallize between 100°-220°C which is its melt temperature.

Packing of Molecules

Polymers obey the principle of close packing and macromolecules must be packed as closely as possible in the crystalline lattice. There are several possibilities for the formation of close packings of polymer chains.

The first crystalline structure is built by close packing of spheres which is exhibited by globular proteins. The formation of such crystalline structures are possible because all the spheres are equal in size, *e.g.*, edestion catalase, iodobenzoyl glycogen.

The second possibility of close packing is the packing of helical macromolecules. The convexities of one helix fit into the concavity of other, *e.g.*, amylose, poly (hexamethylure adipamide), polypeptides.

The third is the packing of long extended chains, *e.g.*, polyamide, polyurethanes, etc. The essential requirement is that the side chain substituents must not hinder regular arrangement of adjacent chains.

Close packing of chain molecules can only be accomplished with flexible chains capable of shifting in parts. Introduction of polar groups gives rise to two opposite effects. On one hand, intramolecular attraction increases, favouring close packing, while on the other hand, the thermodynamic flexibility (potential energy difference between two positions of the chain) of the chain decreases. The chain becomes more rigid. The rate of close packing depends on these two factors. However, with polar groups like –OH or –CN, the attraction effect prevails inspite the fact that the chains are rigid.

Since rigid chains closely pack in the extended state, such polymers are highly oriented, *e.g.*, poly (vinylalchol), polyacrylo butyrate. The chains of isotactic polypropylene are highly flexible and form crystalline lattice, but rigid cellulose chains cannot form such lattices under any condition.

Polymer Crystal Cells

The crystalline cell of polymers consists of parts of chains located at definite distances a, b, and c and angles α, β and γ. They do not differ at all from cells formed by low molecular weight compounds and obey all the rules of symmetry.

Polyethylene crystals form orthorhombic unit cells with a = 7.40Å, b = 4.93Å and c = 2.53Å and density equal to 1.0 g/cm^3. All the carbon atoms in the polyethylene cell are arranged on the same plane forming a zigzag plane. In the polyethylene unit cell, chains are arranged along four edges and in the centre of the cell.

In the case of chains having bulky side substituents helical crystals are formed, *e.g.*, polypeptides, polypropylene divinyl and olefin polymers. One helical loop may contain a different number of monomeric units depending on the nature of a polymer. Natta synthesized 1,2 polybutadiene and similar results were experienced with isotactic polystryrine. Polymers like sulphur or phosphorus are characterized by poly-morphism. Polypropylene may form crystals belonging to monoclinic, hexagonal and triclinic types of symmetry. They are typical first order phase transitions. These transitions between various polymorphous formations take place either upon a change in temperature or under the action of load which brings a change in the parameters of a crystalline cell. As a matter of fact most of the polymers can be crystallized under stretched conditions.

The morphology of a crystalline polymer is determined by the mutual arrangement of the unit cells within the limits of the crystalline state of a body.

Globular Crystals

In globular crystals, macromolecules in coiled conformation form the lattice points. These are found in biopolymers, as a very high degree of uniformity of macromolecules is a pre-requisite for it, *e.g.*, tobacco mosaic virus synthetic polymers do not form them.

Spherulites

Spherulites are obtained when polymers crystallize from melts or concentrated solutions. They are made of crystallites which grow radially from one common centre. Macromolecular chains make at least an angle of 60° with the radius, *i.e.*, they are arranged tangentially to the spherulite radius Spherulites possess anisotropic properties. Their indices of refraction of light are different in radial or tangential directions. The orientation of the crystallographic axes continuously changes along the angular coordinates, A spherulite is called positive when the refractive index in the radial direction is greater than in the tangential one; otherwise it will be a negative spherulite.

In radial Spherulites one of the axis of the crystalline lattice retains a constant direction along all the radial directions. Ring type Spherulites are built of lamellar crystals whose orientation continuously changes along the spherulite radius.

Depending on the degree of cooling, the same polymer may form different types of spherulitis. With low degrees of super cooling ring type Spherulites are obtained while radial Spherulites are obtained by high ones.

The formation of Spherulites passes through several stages. First the crystal nuclei are formed which are scattered throughout the volume of the sample and then independent growth of discrete fibrillar or lamellar crystalline structure grow at the same time in all directions. Then encountering one another they grow into regions filling the entire volumes of the body. Then their boundaries are distorted and they assume the form of a polyhedron.

The structural units in spherulite are bundles of macromolecules made out of chains in an extended conformation. This is possible when several macromolecules try to crystallize simultaneously from both the ends. Such interstructural bonds combine to form separate crystallites inside fibrillar or lamellar crystals.

Oriented State of Polymers

Polymers because of great anisotropy of the dimensions of macromolecules exist in oriented state. In the extreme case all the macromolecules will be oriented parallel to each other, but in reality this can happen on the average as the flexible chains do retain some fold conformations. In the oriented state the difference in amorphous and crystalline polymer vanishes.

Peterlin suggested that in oriented polymers microfibrils of diameter 100-200Å clearly divide crystalline blocks of chains with fold conformation. These chains are divided by defective layers and a large number of intra- and smaller number of interfibrillar chains. The microfibrils continue into larger fibrils to create the macrostructure of oriented bodies of the size 1000Å.

These polymers are important for producing the fibres and films. In the first case an uniaxial orientation takes place while in the second a planar orientation takes place.

Liquid Crystals

Certain substances that contain rod-shaped molecules can exist between the crystalline and liquid state. The forces between these molecules are greatest when they-lie parallel and the crystals of such substances consist of parallel arrays of molecules stacked side by side when the crystal is heated, a point is reached at which the thermal energy is sufficient to break down the crystal lattice but not the parallel arrangement of the molecules. This results into a liquid crystal when the molecules are free to move provided they all remain parallel. At higher temperature the liquid crystal melts to a liquid. Such crystals are called *nematic.*

There are other liquid crystals where the molecules are not only parallel but their ends are aligned. Such crystals are known as *sematic.* In this case the crystals contain layers where the parallel molecules are free to move about in its layer but not in other layers. These crystals form drops in which the surface forms sharp steps with flat risers in between.

The intermolecular attraction in liquid crystals do not act equally in all directions with the result that all the degrees of freedom are not liberated. In a nematic liquid crystal the molecule can rotate only about its long axis and two rotational degrees of freedom are lost. In sematic liquid crystal the molecules are free to move only in the two directions and some translation degree of freedom is lost.

ELEMENTS

Elements are the basic constituents of all matter. An element is the simplest form of matter, and therefore, cannot be split into simpler substances by any chemical or physical method. In terms of the atomic theory, an element is composed of atoms of the same kind. There are 115 elements known to us, out of which 92 are naturally-occuring while the rest have been prepared artificially. A complete list of the elements is given on the inner-side of the back cover of the book. These elements are widely distributed in the earth's crust in the free as well as in the combined forms.

Classification of Elements

Elements are further classified into metals, non-metals and metalloids. Listed below are some common metals, non-metals and metalloids.

Table 2.4

Metals	*Non-metals*	*Metalloids*
Gold	Hydrogen	Boron
Silver	Oxygen	Silicon
Copper	Nitrogen	Germanium
Iron	Sulphur	Antimony
Mercury	Carbon	Arsenic
Zinc	Phosphorus	

Metals : Elements (except hydrogen) which form positive ions by losing electrons during chemical reactions are called metals. Thus, metals are electropositive elements.

Metals (except mercury) are solid under normal conditions. Metals are very good conductors of electricity and heat. Metals are malleable (can be beaten into very thin sheets), and ductile (can be drawn into wires). Gold, silver, copper, tin, lead, iron and mercury are the metals which were known to the ancient people.

Non-metals : Elements which tend to gain electrons to form anions during chemical reactions are called non-metals. Thus, non-metals are electronegative elements. Non-metals (except graphite) are poor conductors of heat and do not conduct electricity. Non-metals may be gaseous, liquid or solid. Solid non-metals are brittle. Elements such as, oxygen, nitrogen, carbon, phosphorus, sulphur are typical non-metals.

Metalloids : The elements which behave like both metals and non-metals are called metalloids. Boron, silicon, germanium are typical metalloids.

COMPOUNDS

A compound is a pure substance made up of two or more elements combined in a definite proportion by mass. For example, water (H_2O) is a chemical compound made up of hydrogen and oxygen combined in the ratio 2 : 1 by volume or 1 : 8 by mass.

Compounds are commonly called chemical compounds, because these are formed due to the chemical combination of the combining elements.

It may be remembered that any two (or more than two) elements may combine in more than one way to give more than one compounds. For example, the elements carbon and oxygen react to form carbon monoxide (CO) and carbon dioxide (CO_2) gases.

Characteristics of a Chemical Compound

All chemical compounds have the following characteristics.

(i) The chemical compounds are obtained by the chemical union of two or more elements in a definite proportion.

(ii) Compounds are homogeneous.

(iii) The chemical and physical properties of a compound are entirely different from those of the component elements. For example, the properties of water are altogether .different from those of hydrogen and oxygen.

Hydrogen	*Oxygen*	*Water*
Combustible gas	A gas which supports combustion	A liquid which is neither combustible nor supporter of combustion.

(iv) During the formation of a compound, energy in the form of heat, light or electricity is either evolved or absorbed. For example, when coal is burnt, heat and light energies are evolved.

Classification of Compounds

Based on the origin or chemical constitution, the compounds may be broadly classified into the following two types.

(a) *Inorganic compounds* : The compounds which are obtained from non-living sources such as minerals are called inorganic compounds. Common salt, limestone, marble, gypsum are inorganic compounds.

(b) *Organic compounds* : The compounds which occur in plants or animals are called organic compounds. Organic compounds can also be synthesized in laboratories. For example, carbohydrates, proteins, oils, fats, waxes, etc., are common organic compounds.

The compounds may be further classified into acids, bases, and salts.

MIXTURES

A mixture is the system in which two or more homogeneous substances (elements or compounds) are simply mixed together in any proportion. All mixtures (except solutions) are heterogeneous. The components of any mixture can be separated by simple laboratory methods, such as, filtration, distillation, sublimation etc.

Mixtures of Two Types

(a) Homogeneous mixtures

(b) Heterogeneous mixtures.

Depending upon the physical states (gas, liquid or solid) of the components of any mixture, the following types of mixtures are possible.

Table 2.5

Type of Mixture	*Example*	
	Homogeneous	*Heterogeneous*
1. Gas in Gas	Air, (O_2 + H_2)	–
2. Gas in liquid	Aerated water, (H_2O + CO_2)	–
3. Gas in solid	Hydrogen in palladium	–
4. Liquid in liquid	Ethyl alcohol + Water	Water + Oil
5. Liquid in solid	Mercury in amalgamated zinc	–
6. Solid in liquid	Sugar in water	Chalk in water Dust (fine sand) in water
7. Solid in solid	Alloys, *e.g.*, brass	Mixtures such as, sand + iron fillings; sand + ammonium chloride etc.

SEPARATION OF THE CONSTITUENTS OF A MIXTURE: SOME SIMPLE LABORATORY TECHNIQUES

Mixtures contain two or more substances in any proportion. One of the most interesting aspect of chemistry is to obtain substances in pure state from mixtures. A substance can be purified by taking advantage of its chemical and physical properties. *The constituents of any mixture can be separated by taking advantage of the differences in their physical properties such as, solubility, particle size, effect of heat, etc.*

The constituents of any mixture are separated by using certain common laboratory techniques. Techniques commonly used for separating the constituents of mixtures are described below.

Sedimentation and Decantation

This method is generally used for separating the constituents of a mixture in which one component is a liquid, and the other is in the form of coarse particles of a solid heavier than the liquid. For example, muddy river water.

Principle : This method is based on the effect of gravity. The coarse particles of solid being heavier than water, settle down, and the clear upper layer of the liquid is gently poured out into another container.

Process : To separate the coarse particles of a solid from a liquid (say, muddy river water), take the mixture in a container and allow it to stand for sometime. The solid particles settle down to the bottom of the vessel due to gravity.

Settling down of coarse particles under the influence of gravity is called sedimentation. During sedimentation, heavier/bigger particles settle down faster than the finer particles. Settling down of the particles leaves the upper layers of liquid clearer.

The clear supernatent liquid (upper layer of liquid) is carefully transferred to another beaker. *This process of mechanically transferring of the clear liquid without disturbing the settled solid particles, is called decantation.*

A laboratory set up for sedimentation, and decantation is shown in Fig. 1.4.

Very fine particles which do not settle down cannot be removed by sedimentation and decantation. Sometimes, to quicken the settling (sedimentation) process, a small quantity of alum is added to the muddy water. Al^{3+} ions cause the coagulation of the fine clay particles, and make their settling faster. Al^{3+} ions in water/aq. solution undergo hydrolysis to form $Al(OH)_3$.

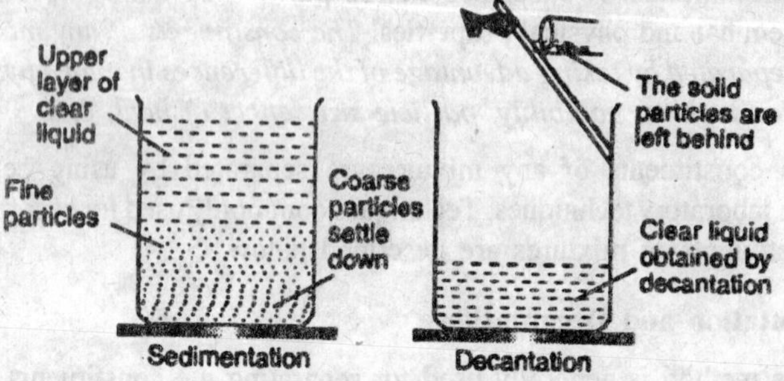

Fig. 2.6 : Separation of the coarse particles of a solid from a liquid by sedimentation and decantation.

$Al(OH)_3$ has strong tendency for adsorption. So, these precipitate of $Al(OH)_3$ adsorb finer clay particles on them and settle down.

Filtration

Filtration is used for separating the insoluble solid component of a mixture from the liquid completely. In chemistry, this method is used for separating the precipitate (solid phase) from any solution.

Principle : This process is based on the fact that solvent molecules and the molecules/ions present in the solution can pass through the porous membranes (such as, filter paper), while the suspended particles cannot. As a result, such particles are retained on the porous membrane, and separated from the liquid phase.

Process : The solution containing suspended impurities is made to pass through a porous membrane, such as, filter paper, filter cloth, etc. The solvent or solution containing dissolved substances passes through the porous membrane. This is called filtrate. The insoluble suspended particles remain on the porous membrane. This solid component left on the porous membrane is termed residue. In laboratory filtration is carried out by using filter paper and a funnel. A filter paper cone is placed in a glass funnel. The funnel containing this paper-cone is then supported on a funnel stand, and the impure solution is transferred to the paper cone with the help of a glass rod. Be careful not to fill more than two-third of the paper cone at any time. The filtrate may be collected in a small beaker. The insoluble impurities which do not pass through the filter paper remain on it. A typical experimental set up for filtration using a filter paper is shown in Fig. 2.7.

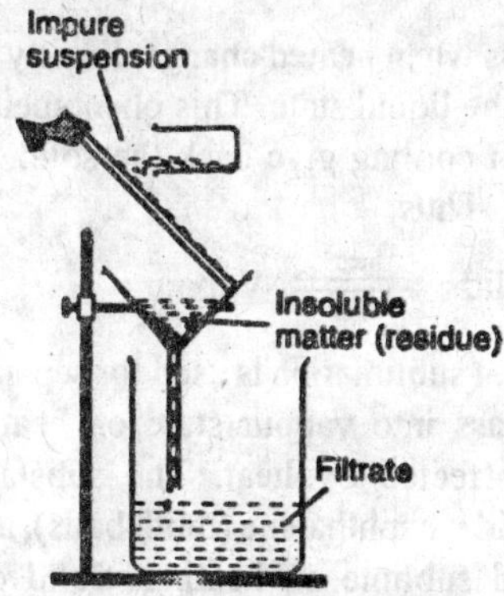

Fig. 2.7 : Separation by filtration

When the suspended impurities are very fine, then the addition of a small amount of alum to the suspension makes the filtration more convenient and faster. This happens because alum causes coagulation of the finer particles. The bigger aggregates do not clog the pores of the filter paper, and thus permit quicker filtration.

Evaporation

Evaporation is used for recovering the soluble solid solute from the solution. In this method, the solvent is lost into the surroundings.

Principle : All liquids evaporate at all temperatures. Evaporation becomes faster at higher temperatures.

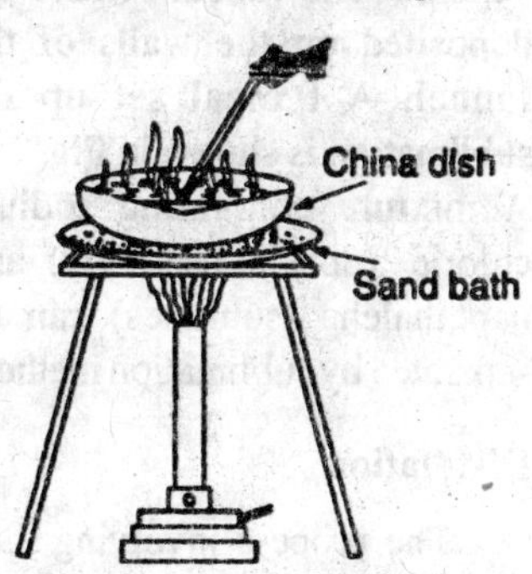

Fig. 2.8 : Evaporation of a solution on a sand bath, to remove the solvent.

Process : The solution containing the dissolved solute is taken in a china dish, and heated on a low flame. When most of the solvent is gone, the solution becomes thicker. At this stage, the flame is reduced, and slowly the semi-solid mass in the china dish is heated to dryness.

Solution containing soluble salt $\xrightarrow{\text{heated}}$ Solid dry salt left behind in the china-dish + Solvent vapour escape to the atmosphere

Sublimation

Certain solids when heated change directly into vapour state without passing through the liquid state. This phenomenon is called sublimation. These vapours on cooling give back the solid. The solid so obtained is called sublimate. Thus,

$$\text{Solid} \underset{\text{cool}}{\overset{\text{heat}}{\rightleftharpoons}} \text{Vapour}$$

The process of sublimation is used for separating the solid substances which directly pass into vapour state on heating from the substances which are not affected by heat. The substances such as camphor, ammonium chloride, naphthalene (moth balls), anthracene, iodine, indigo and benzoic acid sublime on heating. Solid carbon dioxide ($CO_2(s)$) under normal conditions also sublimes. At room temperature and normal atmospheric pressure, solid carbon dioxide changes into CO_2 gas without forming the liquid CO_2. It is due to this that solid carbon dioxide is called as dry ice.

Process : In laboratory, sublimation is carried out by heating the crude sample or a mixture in a china dish. The china dish is covered with an inverted glass funnel to collect the vapours. The vapour formed get deposited on the walls of the funnel. A typical set up for sublimation is shown in Fig. 2.9. A mixture containing sodium chloride (does not sublime), and naphthalene (sublimes) can be separated by sublimation method.

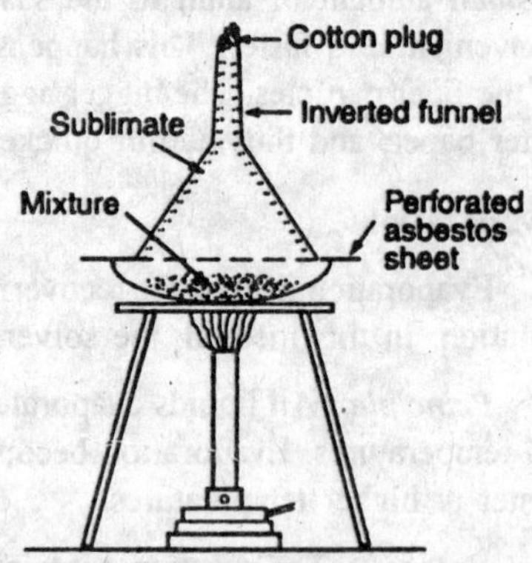

Fig. 2.9 : An experimental set up for sublimation.

Distillation

The process involving the vaporisation of a liquid on heating and condensation back to the liquid on cooling is called distillation.

Distillation is Used

(a) for separating the components of a solution containing a soluble substance, *e.g.*, for separating sodium chloride (salt) and water from a solution of salt in water.

(b) for separating the liquid components of a liquid mixture when the liquids have different boiling points.

Depending upon the difference in the boiling points of the liquids different types of distillations are used. These are described below.

(i) *Simple distillation :* This technique is used for separating liquids having boiling points differing by 10-20 degrees. The liquid having lower boiling point distils over and the other liquid is left behind. In this process, vaporisation and condensation occur side by side.

In laboratory, simple distillation is carried out by using an experiment set up shown in Fig. 2.10.

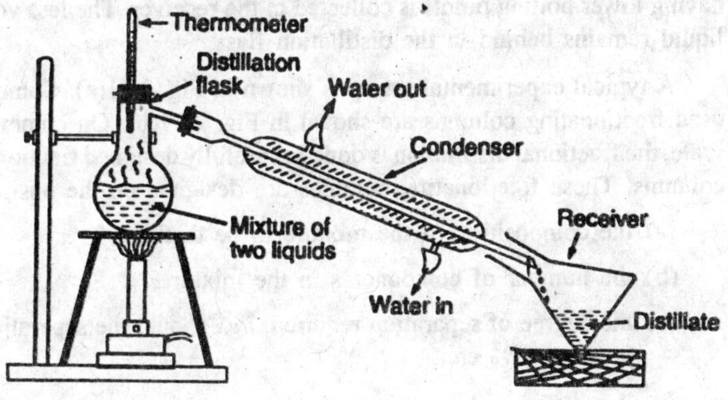

Fig. 2.10 : An experimental set up for simple distillation.

The liquid mixture is taken in a distillation flask fitted with a thermometer and a condenser. The distillation flask is heated on a sand bath or on a wire gauge. *The more volatile liquid* (liquid having lower boiling point), *boils first and its vapours leave the flask from the outlet near at the top.* These vapours then pass through the condenser, and get condensed to liquid. The condensed liquid (called distillate) is collected in a receiver. The less-volatile liquid (liquid having higher boiling point) is left behind in the distillation flask.

To avoid bumping of liquid, a few glass beads or porcelain pieces are placed in the distillation flask.

(ii) *Fractional distillation*: When the boiling points of the two liquids in any mixture do not differ much, they boil within a narrow range of temperature. In such cases, ordinary distillation does not give complete separation. Instead, a modified version of this technique called fractional distillation is used. In fractional distillation, vapours from the boiling liquid mixture are made to pass through a glass fractionating column. A fractionating column is a glass column packed with glass beads or a specially designed column. The vapours of the low volatile liquid (high boiling component) get condensed in this column, and return back to the distillation flask, while the vapours of the high volatile liquid leave the column from the exit near the top.

These vapours leaving the column enter the condenser, and get condensed. The condensate consisting of pure more volatile liquid (liquid having lower boiling point) is collected in the receiver. The less volatile liquid remains behind in the distillation flask.

A typical experimental set up is shown in Fig. 2.11(a). Commonly used fractionating columns are shown in Fig. 2.11(b). On commercial scale, the fractional distillation is done in carefully designed fractionating columns. These fractionating columns are designed on the basis of

(a) the composition of the mixture to be distilled,

(b) the number of components in the mixture,

(c) the degree of separation required, *i.e.*, should the separation be 90% or 95% etc.

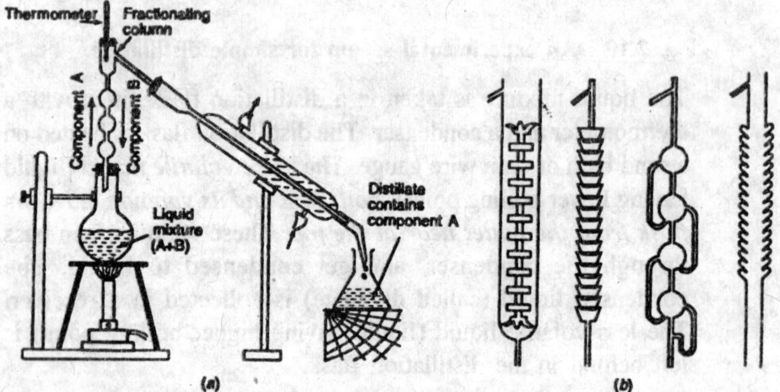

Fig. 2.11 : (a) An experimental set up for functional distillation. The liquid A is more volatile, and the liquid B is less volatile, [b) The commonly used fractionating columns.

One typical example of applications of this technique on commercial scale is the fractional distillation (refining) of crude petroleum to get different fractions, such as, gasoline, kerosene, lubricating oil etc.

Crystallisation

Crystallisation is used for obtaining a solid compound in pure and geometrical form.

Principle : The impure sample of a solid substance is dissolved in a suitable solvent to prepare its saturated solution at a temperature slightly higher than the room temperature. When such a solution is cooled, the substance reappears in the form of well-shaped crystals. The solution left behind is called mother liquor. All the impurities are left behind in the mother liquor.

Process : A nearly-saturated solution of an impure substance is prepared in a hot solvent. The solution so prepared is filtered quickly. The filtrate is allowed to cool slowly in a china dish. Crystals of pure substance formed are removed with the help of a spatula, and dried by pressing them between the folds of filter papers, and finally in a desiccator.

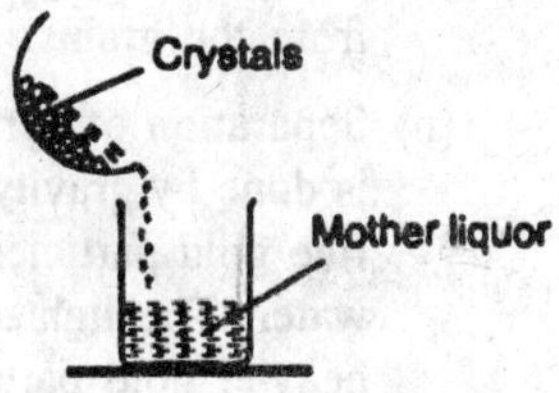

Fig. 2.12 : Crystallisation: separating crystals from the mother liquor.

Magnetic Separation Method

A magnetic component of a mixture, *e.g.*, iron can be separated with the help of a magnet. For example, the mixture containing iron fillings (magnetic substance) is placed in a glass plate, and a magnet is moved over the mixture.

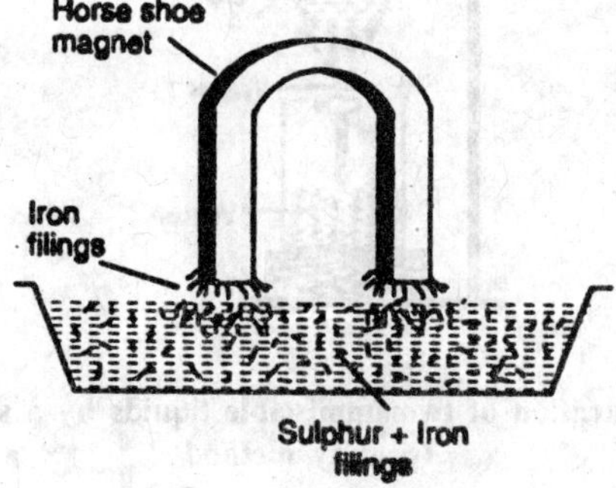

Fig. 2.13 : Separation of a magnetic substance by a magnet.

The magnetic material (*e.g.*, iron fillings) gets attached to the magnet. The process is repeated number of times till the magnetic material is completely separated from the mixture (Fig. 2.13).

Gravity Method

Gravity method takes advantage of difference in the density of the two components of a mixture. The actual operation depends upon the type of mixture and its components. Some typical cases are described below.

(i) Chaff is removed from the grains with the help of wind (this is called winnowing). When the grain and chaff is made to fall from a height in the blowing wind. The heavier grains fall straight to the ground, while the lighter chaff falls a little away from the grains.

(ii) Separation of very fine gold particles from the river bed sand is done by gravity method. In this method, the sand containing fine gold particles is repeatedly washed in a pan with flowing water. The lighter sand particles get washed away, while the heavier gold particles settle at the bottom of the pan.

(iii) Two immiscible liquids having different densities, *e.g.*, oil and water, can be separated with the help of a separating funnel. The heavier liquid forms the lower layer. The lower layer is separated by opening the stopcock, and collecting it in a clean and dry beaker. A typical set up is shown in Fig. 2.14.

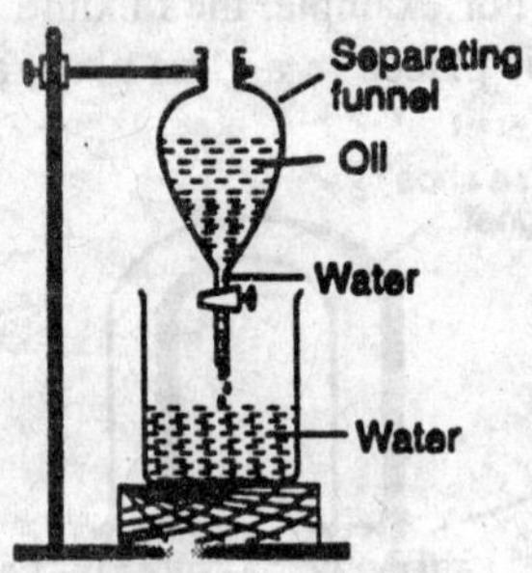

Fig. 2.14 : Separation of two immiscible liquids by a separating funnel (gravity method).

Solvent Extraction Method

Sometimes, a valuable substance is present in any mixture in very small amounts. Separation of this substance by commonly used techniques is not always possible. In such cases another technique called solvent extraction may be used. The actual experimental procedure depends upon the nature of the mixture from which the substance is to be extracted. Two typical cases are described below.

When the mixture containing the substance is a solution (a liquid): If the mixture containing the substance to be extracted is a solution (liquid phase), then a suitable solvent is chosen. The chosen solvent should satisfy the following conditions.

(i) the solvent should be immiscible with the liquid mixture.

(ii) the desired product (the substance to be recovered) should be highly soluble in the chosen solvent.

(iii) the desired product should not get chemically affected by the solvent.

After selecting the solvent, the mixture is shaken with a small quantity of the solvent in a separating funnel, and allowed to rest for sometime. The two layers are separated by opening the stopcock. The desired product is then recovered by evaporating the solvent. The solvent is recovered by condensing the vapour of the solvent. The recovered solvent is used again.

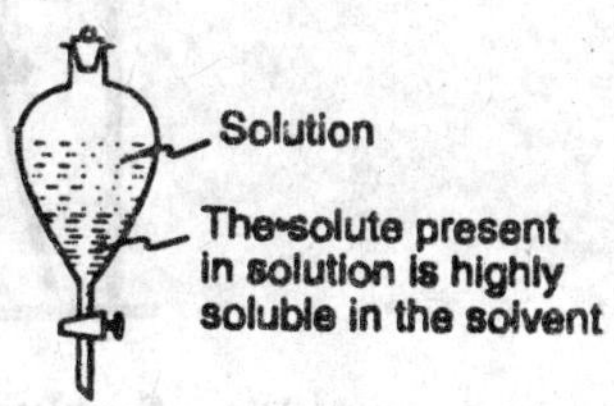

Fig. 1.13 : solvent extraction by using a separating funnel.

A typical example is the separation of iodine from aqueous solutions by using carbon tetrachloride as solvent. The solubility of iodine in water is very low. Iodine is highly soluble in carbon tetrachloride. So, when an aqueous solution containing a small quantity of iodine is shaken with carbon tetrachloride in a separating funnel, iodine gets extracted by carbon tetrachloride. Carbon tetrachloride being heavier than water forms a lower layer. This lower layer can be removed by opening the stopcock of the separating funnel. When distilled, carbon tetrachloride vaporises leaving behind iodine. The vapour of carbon tetrachloride is condensed to get back the solvent (carbon tetrachloride).

When the substance is present in a solid substrate: Certain valuable products occur in nature in solid substrates, *e.g.*, the soluble dyes and pigments or pleasant smelling compounds occur in plant leaves and roots. To extract these compounds, the solid substrate (leaves, flowers or root) are crushed. The crushed material is then repeatedly extracted with hot solvent in a Soxhiet apparatus shown in Fig. 2.16.

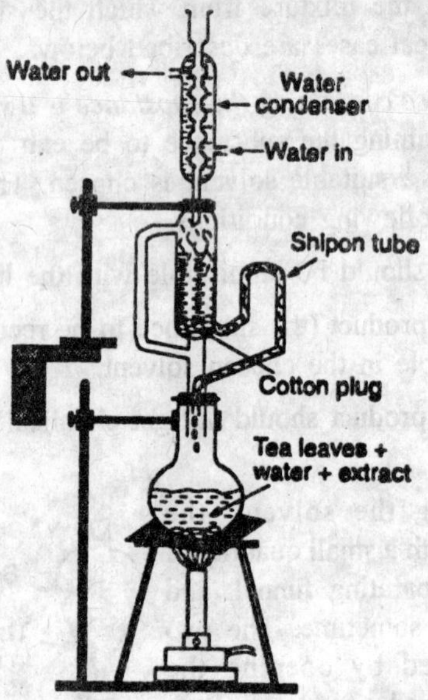

Fig. 2.16 : Soxhiet apparatus for solvent extraction.

The crushed sample in placed in the central wide tube, and the solvent in the flask is boiled. The vapours of the solvent are made to condense on the crushed material. These droplets of the condensate, dissolve the desired product and the dilute solution so produced returns back to the flask through the siphon tube. The process continues and more and more of the product gets extracted into the solution. The desired compound in the solid form can be obtained by either crystallisation or by precipitation from the solution.

Typical example of the solvent extraction technique is the preparation of tea-water (soluble fraction having colour and flavour in the tea leaves gets dissolved in boiling water).

Extraction of essential oils of flowers and leaves (*e.g.*, Rooh Gulab) is also based on the solvent extraction technique.

Chromatography

Chromatography is a valuable experimental technique for the separation, purification, and identification of the constituents of a mixture. This method is based on the differential adsorption of the various components of a mixture on a suitable adsorbent. Suitable adsorbents are, magnesium oxide, alumina, cellulose paper, silica gel etc. The adsorbent is termed as fixed or *stationary phase* and the liquid in which substance is dissolved is which substance is dissolved is termed as *mobile* or *moving phase.* Depending upon the nature of the two phases there are various types of Chromatography. Here, we describe the most commonly used type of Chromatography, known as column Chromatography.

Column Chromatography : This is also called adsorption Chromatography. In this method, an adsorbent alumina (Al_2O_3) is packed in a column (column is a burette-like glass tube). The column of Al_2O_3 acts as the stationary phase.

The mixture to be separated is dissolved in a suitable solvent and the solution is poured on top of the column. The mixture moves down, and the different components of the mixture get adsorbed on the alumina surface. Some may get adsorbed strongly, while some other may be adsorbed weakly. The adsorbed components are then eluted out by using a suitable solvent.

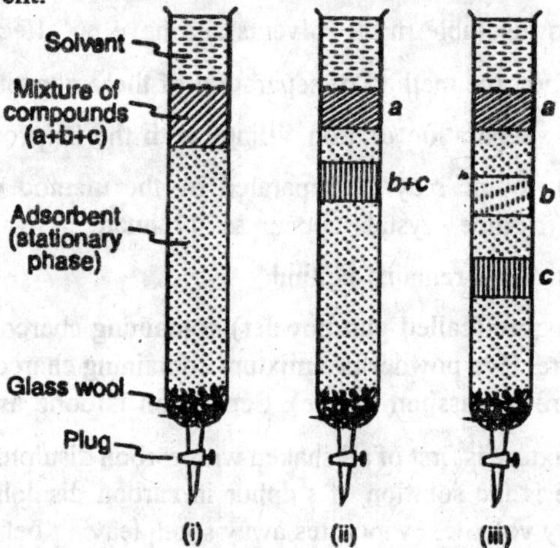

Fig. 2.17 : Diagrammatic representation of column Chromatography [(i), (ii) and (iii) show the progressive separation of the components a, b and c of a mixture with a solvent (eluent)].

This solvent acts as the mobile phase. The weakly adsorbed component will be eluted out (washed out by the mobile phase) more rapidly than the more strongly adsorbed component. Thus, different components of the mixture come out of the column one by one. Such a progressive separation of a mixture is shown in Fig. 2.17.

The other adsorbents which can be used in the column are, magnesium oxide, silica gel, activated animal charcoal etc.

Separation of the Components of Some Typical Mixtures

Separation of the components of some typical mixtures is described below.

(i) Mixture containing powdered sulphur and iron fillings. The mixture is a heterogeneous mixture containing a magnetic substance (iron). So, the iron can be separated from sulphur by moving a magnet several times over the mixture.

(ii) Mixture containing iodine, sand and iron fillings. The three components of the mixture are characterized by their properties, viz;

Iron is a magnetic substance.

Iodine undergoes sublimation on heating.

Sand is insoluble in all solvents and have no effect of heat on it.

Therefore, the method of separation of these constituents involves,

(a) Separation of iron fillings with the help of a magnet.

(b) Iodine may be separated by the method of sublimation. Iodine crystals appear as sublimate.

(c) Sand remains behind.

(iii) Mixture (called gun powder) containing charcoal, sulphur and nitre. Gun powder is a mixture containing charcoal, sulphur and nitre (potassium nitrate). Separation is done as follows.

The mixture is first of all shaken with carbon disulphide and filtered. The filtrate is the solution of sulphur in carbon disulphide (CS_2). CS_2 being highly volatile, evaporates away soon, leaving behind the crystals of sulphur.

The residue on the filter paper containing charcoal and nitre is then shaken with a small quantity of hot (nearly boiling) water, and filtered.

The filtrate contains nitre. Nitre crystals can be obtained by evaporating the filtrate to dryness. The charcoal left on the filter paper is washed and dried in hot air.

Flow sheet of the methods used in separating the constituents of gun powder is given below.

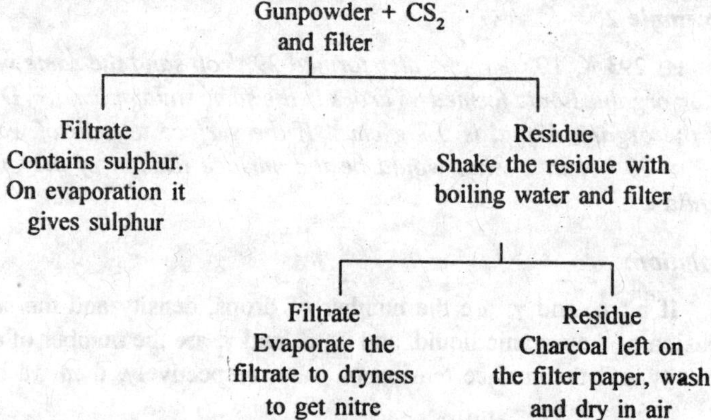

(iv) Mixture containing two immiscible liquids. Two immiscible liquids can be separated with the help of a separating funnel. The mixture containing such liquids is placed in a separating funnel and allowed to stand for some time, the heavier liquid forms the lower layer and can be taken out by opening the stopcock of the funnel.

Some typical pairs of immiscible liquids are:

(i) Carbon tetrachloride + Water (carbon tetrachloride is heavier than water)

(ii) Oil + Water (oil is lighter than water)

(iii) Carbon disulphide + Water (carbon disulphidè is heavier than water).

SOLVED EXAMPLES

Example 1:

At 20°C, water formed 29 drops when flowing through the capillary of a stalagmometer, while an equal volume of ether formed 86 drops. If the densities of water and ether are 0.997 and 0.70 gm per cc respectively, find the surface tension of ether if that of water is 72.8 dynes/cm.

Solution:

We have $\gamma_2 = \frac{n_1}{n_2} \cdot \frac{\rho_2}{\rho_1} \cdot \gamma_1$

$= \frac{29}{86} \times \frac{0.07}{0.997} \times 718 = 17.24$ dynes/cm

Example 2:

At 293 K, $10^{-2} dm^3$ of water formed 29 drop sand the same volume of an organic liquid formed 86 drops in the same stalagmometer. Density of the organic liquid is 0.7 g cm^{-3}. If the surface tension of water is 7.2×10^{-2} N m^{-1}, what would be the surface tension of the organic liquid ?

Solution:

If n_1, ρ, and γ_1 are the number of drops, density and the surface tension of the organic liquid, and n_2, d_2 and γ_2 are the number of drops, density and the surface tension of water respectively, then we have

$$\frac{\gamma_1}{\gamma_2} = \frac{n_1\rho_1}{n_1\rho_2}$$

or $$\gamma_1 = \frac{\gamma_2 n_2 \rho_1}{n_1 \rho_2}$$

Since $n_2 = 29$ $\quad$ $n_1 = 86$

$\rho_2 = 1$ g cm^{-3} $\quad$ $\rho_1 = 0.70$ g cm^{-3}

$= 1 \times 10^3 gm^{-3}$ $\quad$ $= 7.0 \times 10^2$ gm^{-3}

$\gamma_2 = 7.2 \times 100^{-2} Nm^{-1}$

$\gamma_1 = ?$

Therefore

$$\gamma_1 = \frac{29(7.0 \times 10^2 \text{ kg m}^{-3})(7.2 \times 10^{-2} \text{ N m}^{-1})}{86\,(10^3 \text{ kg m}^{-3})}$$

$= 1.726 \times 10^{-2}$ m^{-1}

or $= 17.26$ dynes cm^{-1}

Example 3:

Equal volumes of an organic liquid and water gave 55 drops and 35 drops respectively. The densities of water and the organic liquid are

Nm^{-1}. Calculate the surface tension of the organic liquid. How many times a water drop is heavier than a drop of the organic liquid ?

Solution:

$$\frac{\gamma_1}{\gamma_2} = \frac{n_2 d_1}{n_1 d_2}$$

or $$\gamma_1 = \frac{\gamma_1 n_2 d_1}{n_1 d_1}$$

Since

$n_1 = 55$ $\quad$ $n_2 = 35$

$d_1 = 0.80 g\ cm^{-3}$ $\quad$ $d_2 = 0.996 g\ cm^{-3}$

$= 8.0 \times 10^2\ kg\ m^{-3}$ $\quad$ $= 9.96 \times 10^2\ kg\ m^{-3}$

$\gamma_1 = ?$ $\quad$ $\gamma_2 = 7.2 \times 10^{-2}\ Nm^{-1}$

Therefore

$$\gamma_1 = \frac{(7.2 \times 10^{-2}\ Nm^{-1})(35)\ (8.0 \times 10^2\ kg m^{-3})}{(55)\ (9.96 \times kg m^{-3})}$$

$$= 2.63 \times 10^{-2}\ Nm^{-1}$$

Now, $$\frac{m_{water}}{m_{organic\ liquid}} = \frac{\gamma_{water}}{\gamma_{organic\ liquid}}$$

$$= \frac{7.2 \times 10^2\ Nm^{-1}}{2.63 \times 10^{-2}\ Nm^{-1}} = 2.74.$$

Example 4:

Water requires 120.5 seconds to flow through a viscometer and the same volume of acetone requires 49.5 seconds. If the densities of water and acetone at 293 K are 9.982×10^2 kg m^{-3} and 7.92×10^2 kgm^{-3} respectively and the viscosity of water at 293 K is 10.05 pascal second, calculate the viscosity of acetone at 293 K.

Solution:

Since, $$\frac{\eta_{water}}{\eta_{Acetone}} = \frac{t_1 \rho_1}{t_2 \rho_2}$$

$$\eta_{Acetone} = \frac{\eta_{water}\ t_2 \rho_2}{t_1 \rho_2}$$

$$= \frac{(1.005 \times 10^1\ Pas)\ (49.5s)\ (7.92 \times 10^2\ kg m^{-3})}{(12.5s)\ (9.982 \times 10^2\ kg\ m^{-3})}$$

Example 5:

Calculate the pressure of 10 moles of ethane kept in a vessel of volume 4.86 litres at 27°C.

Solution:

(a) Ideal gas equation

$$P = \frac{nRT}{V} = \frac{100 \times 0.082 \times 300}{4.86} = 50.7 \text{ atm}$$

(b) Van der Waals equation

$$P = \frac{nRT}{\upsilon - nb} - \frac{an^2}{\upsilon^2} = \frac{10.0 \times 0.082 \times 300}{4.86 - 10.0 \times 0.0643} - \frac{5.44 \times 10.0^2}{(4.86)^2}$$

$= 58.4 - 23.0 = 35.4$ atm.

(c) Dietericis equation

$$P = \left(\frac{nRT}{V - nb}\right) e - \frac{an}{VRT}$$

$b = 0.708$ 1 $mole^{-1}$; $a = 7.111^2$ atm $mole^{-2}$

$$P = \left(\frac{nRT}{V - nb}\right) e - \frac{an}{VRT} \quad e = 7.11 \times 10/4.86 \times 0.082 \times 300$$

$= 59.3\ e^{-0\text{-}594} = 32.7$

(d) Reduced equation

$$T_R = \frac{T}{T_c} = \frac{300}{191} = 157$$

$$V_r = \frac{V}{V_c} = \frac{4.86}{0.99} = 4.9$$

$$P_r = \frac{P}{P_c} = \frac{P}{45.8}$$

From Eq. 126

$$P = P_c P_r = 45.8 \times .71 = 32.6 \text{ atm}$$

It can also be obtained by finding out the value of P_r from the value of β for methane in Fig. 3.51 but the actual value observed by B.H. Sage *et al.* was 34.0 atm. Thus, marked deviations are observed in almost all the cases showing that all these including the corresponding states relations are only approximations. However, the corresponding states relations have proved a very quick method of finding out the parameters of a non-ideal gas.

Example 6:

A sample of oxygen is collected by displacing water from an inverted tube. The temperature is 25°C, the pressure is 750 mm Hg (barometric pressure) and the volume occupied is 280 cm³. What is the true volume of oxygen at STP?

Solution:

As the gas is collected over water barometer pressure (P_{bar}) does not correspond to the actual pressure of the gas

$$P_{bar} = + PH_2O$$

$$P_{O_2} = P_{bar} - P_{H_2}O$$

$$= (750.0 - 23.8) \text{ mm Hg} = 726.2 \text{ mm Hg.}$$

Volume of gas at STP can be calculated as:

Initial state: $P_1 = 726.2$ mm Hg; $V_1 = 280$ cm³;

$n_1 = n$; $T_1 = 273 + 25 = 298°K$

$P_1V_1 = nRT_1$

Final state: $P_1 = 760$ mm Hg, $V_2 = ?$ $n_2 = n$; $T_2 = 273$ $P_2V_2 = nRT_2$

As nR is the same in the two states, therefore,

$$\frac{P_1V_1}{T_1} = \frac{P_2V_2}{T_2}$$

$$V_2 = \frac{P_1}{P_2} \times \frac{P_2V_2}{T_2}$$

$$= \frac{726.2 \text{ mm Hg}}{760 \text{ mm Hg}} \times \frac{273° K}{298° K} \times 280 \text{ cm}^3$$

$$= 245 \text{ cm}^3.$$

Example 7:

You are given three vessels A, B and C, all at the same temperature, with volumes respectively of 1.20 1,2.63 1 and 3.05 1. Vessel A contains 0.695 g of nitrogen gas at a pressure of 742 mm Hg, vessel B contains 1.10 g argon gas at a pressure of 383 mm Hg, vessel C is completely empty at the start of the experiment. What will the pressure become in vessel C if the contents of A and B are completely transferred to C?

Solution:

Nitrogen was originally present in vessel A (V = 1.2 1) and has a pressure of 742 mm Hg. It is now transferred to vessel C, the volume of which is 3.05 1.

Argon was originally in vessel B (V = 2.63 1) and has a pressure of 383 mm Hg. This is transferred to vessel C(C = 3.05 1).

Now we can treat the two gases separately and find their pressures when present in vessel C. The total pressure in vessel C is equal to the sum of the pressures of nitrogen and oxygen in vessel C.

Gas	Vessel	Initial conditions	Final conditions	Partial pressures (P) in vessel C
				$P_1V_1 = P_2V_2$
Nitrogen	AC	$V_1 = 1.2$ 1	$V_2 = 3.05$ 1	$P_{N_2} = \dfrac{1.21 \times 742 \text{ mm Hg}}{3.05l}$
		$P_1 = 742$ mm Hg	$P_2 = ?$	= 294 mm Hg
Argon	BC	$V_1 = 2.631$	$V_2 = 3.05$ 1	$P_{Ar} = \dfrac{2.631 \times 383 \text{ mm}}{3.051}$
		$P_1 = 383$ Hg	$P_2 = ??$	

Pressure of gaseous mixture in C

$= PN_2 + P_{Ar}$

= 294 mm Hg + 330 mm Hg

= 624 mm Hg.

Example 8:

(a) What is the average volume available to a molecule in a sample of nitrogen gas at STP?

(b) Assuming that N_2 molecule is approximately spherical with an effective diameter of 3.6Å, calculate

(i) The actual volume occupied by a N_2 molecule;

(ii) the percentage of the molar volume that is empty space;

(iii) *The average distance between neighbouring gas molecules at STP.*

Solution:

We know that one mole of an ideal gas at STP (*i.e.*, at 0°C and 1 atm pressure) occupies 22.4 litres.

Therefore, assuming ideal behaviour for nitrogen, one mole of this gas will occupy 22.4 litres or 22,400 cm^3 at 0°C and 1 atm, pressure.

There are 6.02×10^{23} molecules in one mole. Therefore, the number of molecules in 1 cubic centimeter is

$$\frac{6.02 \times 10^{23}}{22,400} = 2.69 \times 10^{19} \text{ molecules per cm}^3$$

and the volume available to a gas molecule is

$$\frac{1}{2.69 \times 10^{19}} = 3.72 \times 10^{-20} \text{ cm}^3 \text{ molecule}^{-1}$$

(b) Volume of a sphere $= \frac{4}{3} \pi r^3$

As diameter for nitrogen is 3.6 Å $= 36 \times 10^{-8}$ cm

$$r = \text{radius of nitrogen molecule} = \frac{3.6 \times 10^{-8}}{2} \text{cm}$$

$$= 1.8 \times 10^{-8} \text{ cm.}$$

$$\text{Volume of a nitrogen molecule} = \frac{4}{3} (1.8 \times 10^{-8})^3$$

$$= 2.44 \times 10^{-23} \text{ cm}^3 \text{ molecule}^{-1}$$

Volume of 1 mole of nitrogen molecules

$$= 2.44 \times 10^{-23} \times 6.02 \times 1023 = 14.7 \text{ cm}^3 \text{ mole}^{-1}$$

We find that the actual volume occupied by 1 mole of molecules is 14.7 cm^3 while the gas molar volume is 22,400 cm^3. Thus, the difference 22,400 – 14.7 = 223,85.3 cm^3 is apparently the empty space.

$$\text{per cent empty space} = \frac{\text{Volume empty}}{\text{Volume available}} \times 100$$

$$= \frac{223,85.3}{224.00} \times 100 = 99.9\%$$

The average distance between any molecule and its nearest neighbour can be obtained from the volume available to a gas molecule: Let us assume that the volume available to a molecule has the shape of sphere of radius R. Hence, the volume available to a molecule is 4/3 R^3 and this is equal to 3.72×10^{-20} cm^3 as found in (a)

$$4/3\ R^3 = 3.72 \times 10^{-20}\ cm^3$$

$$R^3 = \frac{3 \times 3.72 \times 10^{-20}}{4x}$$

$$R = 20.7 \times 10^{-8}\ cm$$

Thus, on an avèrage, the volume available to a gas molecule can be assumed to be the volume of a sphere with $R = 20.7 \times 10^{-8}$ cm^3, and accordingly the average distance between the centres of any two neighbouring molecules in the gas sample will be 2R.

Therefore, the average distance between a gas molecule and its nearest neighbour is $2R = 20.7 \times 10^{-8} = 41.4 \times 10^{-8}$cm = 41.4Å.

Another way of computing the distance between a gas molecule and its nearest neighbour in a sample of gas is by assuming that the gas volume is divided into small cubes of edge-length and the volume of this small cube is the volume, on an average, available to a gas molecule. The length of this small cube (1) will then roughly be equal to the average distance between the gas molecules assuming that the molecules are for most part evenly spaced at the centres of identical cubes. If 1^3 is the volume of each small cube, then the value of 1 can be computed from the relation

$$1^3 = 3.72 \times 10^{-20} cm^3$$

where 3.72×10^{-20} is the volume available to each molecule

or $1 = 33.4 \times 10^{-8}$ cm = 33.4Å

From these calculations, we find that the average distance between two neighbouring molecules at STP is about 33.4×10^{-8} cm or 33.4Å (students should find out why this value is different from that obtained previously). The diameter of the gas molecule is only 3.6Å and the distance between molecules in the gas is, therefore, about ten times their diameter.

Similar calculations for solid (density 1.03 gm/cm^3) and liquid (density 0.8 gm/cm^3) nitrogen show that in these states, the average distance between molecules is about 3.56Å and 3.8Å respectively (which are nearly equal to the molecule diameter of nitrogen).

Hence, in solid or liquid states, individual molecules touch each other and it is difficult to compress the substances in these states unlike in the gaseous state.

Example 9:

Naphthalene on oxidation gives phthalic anhydride, a very important raw material for glyptal type of resins. When it is heated to 30°C, it exerts a vapour pressure of 0.177 mm. Calculate the effusion rate when M = 128.16 g/g.mole.

Solution:

$$0.177 \text{ mm} = 2.36 \times 10^5 \text{ g-cm/sec}^2$$

$$Z_m = \frac{\alpha}{4} P \left(\frac{3M}{RT}\right)^{1/2}$$

$$= \frac{0.92}{4} \times 2.36 \times 10^5 \left(\frac{3 \times 128.16}{8.314 \times 303 \times 10^7}\right)^{1/2}$$

$$= 0.23 \times 2.36 \times 10 \left(\frac{128.16}{8.314 \times 10^7}\right)^{1/2}$$

$$= 6.73 \text{ g/cm}^2 \text{ sec.}$$

Example 10:

Compute the relative rates of diffusion of hydrogen and carbon dioxide. Molecular mass of hydrogen is 2 and that of carbondioxide is 44.

Solution:

$$\frac{r_{H_2}}{r_{CO_2}} = \frac{(M_{CO_2})^{1/2}}{(M_{H_2})^{1/2}} = \left(\frac{44}{4}\right)^{1/2}$$

$$= \frac{6.63}{1.41} = 4.7$$

Hydrogen diffuses 3.7 times faster than carbon dioxide.

Example 11:

The speeds of diffusion of CO_3 and O_3 were found to be as 0.29 and 0.271. What is the molecular mass of O_3 if the molecular mass of CO_2 is 44?

Solution:

$$\frac{r_{O_3}}{r_{CO_2}} = \left(\frac{M_{CO_2}}{M_{O_3}}\right)^{1/2}$$

$$\frac{0.271}{0.29} = \left(\frac{44}{M_{O_3}}\right)^{1/2}$$

Squaring both sides

$$\frac{(0.271)^2}{(0.29)^2} = \frac{44}{M_{O_3}}$$

or $$M_{O_3} = 44 \times \frac{(0.29)^2}{(0.271)^2} = \frac{44 \times 0.29 \times 0.29}{0.271 \times 0.271}$$

$= 50.4$ (Mol. mass of ozone is 50.4).

Example 12:

A sample of 1.0 g of gas contained in a 1.0 litre vessel at 24°C exerts a pressure of 0.836 atm. What would be the pressure of the sample if it was compressed to a volume of 0.9631 and heated to a temperature of 47°C.

Solution:

For the gas in the two states we have

Initial state: $P_1 = 0.836$ atm; $V_1 = 1.01$;

$n_1 = N$; $T_1 = 273 + 24 = 297°K$ $P_1V_1 = nRT_1$

Final state: $P_2 = ?$; $V_2 = 0.9631$; $n_2 = n$;

$T_2 = 273 + 47 = 320°K$ $P_2V_2 = nRT_2$

In this problem n is the same in the two states, therefore,

$$\frac{P_1V_1}{T_1} = \frac{P_2V_2}{T_2}$$

or $$P_2 = \frac{V_1}{V_2} \times \frac{T_2}{T_1} \times P_1$$

Substituting the numerical values

$$P_2 = \frac{1.01}{0.9631} \times \frac{320°K}{320°K} \times 0.863 \text{ atm}$$

$= 0.935$ atm.

Example 13:

A mixture of gases at 760 mm Hg pressure contains 65.0% nitrogen, 15.0% oxygen and 20.0% carbon dioxide by volume. What is the partial pressure of each in mm Hg?

Solution:

A fundamental property of gases is that each component of a gas mixture occupies the entire volume of the mixture. The percentage composition by volume refers to the volume of the separate gases before mixing. Thus, 65 volumes of nitrogen, 15 volumes of oxygen and 20 volumes of carbon dioxide, each at 760 mm Hg pressure are mixed to give 100 volumes of mixture at 760 mm Hg.

For a mixture of ideal gases, Dalton's and Amagat's laws are equivalent. Therefore, in a gaseous mixture both the volume fraction υ_i/υ and the pressure fraction P_i/P of a component equal its mole traction $x_i = n_i/n$.

Now as

$$P_i = \frac{\upsilon_i}{V} P$$

Therefore,

$$P_{N_2} = \frac{65}{100} \times 760 \text{ mm Hg} = 494 \text{ Hg}$$

$$P_{O_2} = \frac{15}{100} \times 760 \text{ mm Hg} = 114 \text{ mm Hg}$$

$$P_{CO_2} = \frac{20}{100} \times 760 \text{ mm Hg} = 152 \text{ mm Hg}$$

$$\text{Total pressure} = P_{N_2} + P_{O_2} + P_{CO_2} = 760 \text{ mm Hg}$$

Dalton's law is put to practical use in laboratory work when the amount of gas evolved in a reaction is to be measured quantitatively. Often these gases are collected over water. It is common practice to equalize the level of water inside and outside the gas collecting bottle to make the pressure of the collected gas equal to the pressure of the atmosphere. This is done in order to measure the volume of gas at atmospheric pressure and temperature. If the gas was pure, one could immediately use the gas law to calculate the number of moles (or the quantity) of gas produced by the reaction. However, under the conditions of the experiment the gas collected contains water vapour in addition to the gas of interest. Hence, the pressure of the collected gas is not the atmospheric pressure (P_{atm}) even though the equality of the liquid levels indicates that the total internal pressure is equal to P_{atm}; for this total pressure, according to Dalton's law is made up of the sum of the partial pressure of the gas in question (P_g) and the vapour pressure exerted by the liquid water (p_1) over which the gas is collected. Therefore, $P_{atm} = P_g + P_1$. Pressure of the water vapour is also known as aqueous tension. The pressure of water vapour is dependant on temperature and its values at different temperatures may be found in the book of tables. Whenever a gas is collected over water the pressure of the gas is obtained by subtracting the aqueous tension from the atmospheric pressure (recorded by barometer).

Example 14:

A given sample of ideal gas occupies a volume of 11.21 at 0.863 atm. If the temperature is kept constant, to what pressure will you have to go to change the volume to 15.01?

Solution:

For the gas in the two states we have

Initial state: $P_1 = 0.863$ atm; $V_1 = 11.21$; $n_1 = n$; $T_1 = T$; $P_1V_1 = nRT$

Final state: $P_2 = ?$; $V_2 = 15.01$; $n_2 = n$; $T_2 = T$; $P_2V_2 = nRT$

In this problem nRT is the same in the two states and so

$$P_1V_1 = P_2V_2$$

$$0.863 \text{ atm} \times 11.21 = P_2 \times 55.01$$

$$P_2 = \frac{0.863 \text{ atm} \times 11.21}{15.01}$$

$$= 0.644 \text{ atm.}$$

Example 15:

250 cm³ of a gas is collected over acetone at –10°C and 85 cm Hg. If the gas weighs 1.34 g and the vapour pressure of acetone at –10° is 39 mm Hg, what is the molecular mass of the gas?

Solution:

By Dalton's law

$$P = P_g + P_{acetone}$$

or $$P_g = P - \rho_{acetone}$$

$$= (850 - 39) \text{ mm Hg} = 811 \text{ mm Hg}$$

$$PV = nRT = \frac{g}{M} RT$$

or $$M = g.\frac{RT}{PV}$$

$$R = 0.0821 \frac{1 \text{ atm}}{\text{mole}^\circ K}$$

$$t = 273 - 10 = 263^\circ K$$

$$V = 250 \text{ cm}^3 = \frac{250}{1000} - 1 = 0.25 \text{ l}$$

$$P = 811 \text{ mm Hg} = \frac{811 \text{ mm Hg}}{760 \text{ mm Hg}} \times 1 \text{ atm} = \frac{811}{760} \text{ atm}$$

$$g = 1.34 \text{ g}$$

Substituting,

$$M = \frac{1.34 \text{ g} \times 0.0832 \frac{1}{\text{mole}^\circ K} \times 263^\circ K}{\frac{811}{760} \text{ atm} \times 0.251}$$

Example 16:

A lighter-than-air balloon is designed to rise to a height of 50 kilometres at which point it will be fully inflated. At that altitude the atmosphere pressure is 1.52 mm Hg and the temperature is –5°C. If the full volume of the balloon is one hundred thousand litres, how many kilograms of helium will be needed to inflate the balloon?

Solution

To calculate the number of kilograms of helium required we shall first calculate the number of moles and then convert these to kilograms.

The balloon has flexible walls, hence, the volume of the balloon varies with the applied pressure and at any height the pressure inside the balloon will be substantially equal to the atmospheric pressure. As the atmospheric pressure at a height of 50 km when the balloon is fully inflated is 1.52 mm Hg, the pressure inside the balloon will also be 1.52 mm Hg.

Hence, for the helium at a height of 50 km, we have,

$$P = 1.52 \text{mm Hg}$$

V = Volume of balloon when fully inflated

= 100,0001

= 1.0×10^{51}

n = ?

$$x = 0.821 \frac{1 \text{ atm}}{\text{mol}^\circ \text{K}}$$

T = 273–5 = 268°K

and $$n = \frac{PV}{RT}$$

Before substituting the given numerical values we have to see that the variables in the equation are expressed in the proper units.

If we choose to use the value of R expressed in 1 – atm/mole°K, then the units of the terms in the equation are of necessity to those that appear in R and any quantity or quantities which do not have those units must be converted before substituting in the gas law.

As R is in –1 –atm/mole –°K, the pressure must be in atm, V in L and Tin kelvin.

Therefore, P = 1.52 mm Hg × 1 – atom/760 mm

Hg = 2.0×10^{-3} atm.

V = 1.0×10^5 1 and r = 268°K

Substituting the numerical values

$$n = \frac{PV}{RT} = \frac{20 \times 10^{-3} \text{ atm} \times 1.0 \times 10^5 \text{ 1}}{0.0821 - \frac{1-\text{atm}}{\text{mole}^\circ \text{K}} \times 269^\circ \text{K}}$$

= 9.090 moles.

Since one mole of helium weighs 4 gm the mass of helium is

$$9.09 \text{ moles} \times \frac{4.00}{1\text{mole}} = 36.36 \text{ gm}$$

= 0.03636 kilograms.

Example 17:

The density of oxygen at 25°C is 1.43 g/1 at 1 atm pressure. At what pressure will oxygen have a density twice this value?

Solution:

To solve this problem we use the relation

$$P = \rho \frac{RT}{M}, \frac{P}{\rho} = \frac{RT}{M}$$

For the gas in the two states, we have

Initial state: P_1 = 1 atm; 1 = 1.43 g/1; T_1 = T; $\frac{P_1}{\rho_1} = \frac{RT}{M}$

Final state: P_2 = ?; ρ_2 = 2 × 1.43 g/1; T_2 = T; $\frac{P_2}{2} = \frac{RT}{M}$

Here $\frac{RT}{M}$ is the same in the two states and so

$$\frac{P_1}{\rho_1} = \frac{P_2}{\rho_2}$$

or $$P_2 = \frac{P_2}{\rho_1} \times P_1$$

Substituting the numerical values

$$P_2 = \frac{2 \times 1.43 \text{ g/l} \times 1 \text{ atm}}{1 \times 1.43 \text{ g/l}} = 2 \text{ atm.}$$

Example 18:

A sample of gas weighing 5.75g occupies a volume of 3.41 at 50° and 715 mm Hg pressure. What is the molecular mass of the gas?

Solution:

To solve this problem, we use the equation

$$P = \frac{gRT}{VM}$$

or $$M = \frac{gRT}{PV}$$

In this problem, we have

g = 5.75gm

$$R = 0.0821 \frac{1 \text{ atm}}{\text{mole}^\circ \text{K}}$$

T = 273 + 50 = 323 °K the pressure used must be in atm and V in 1

$P = 7.15\text{mm Hg}$

As R is $\frac{1-\text{atm}}{\text{mole}-°\text{K}}$

$P = 715\text{mm Hg}$

$= 715\text{mm Hg} \times \frac{1 \text{ atm}}{760 \text{ mm Hg}}$

$= \frac{715}{760} \text{ atm}$

$V = 3.41$

Substituting the values,

$$M = \frac{5.5\text{g} \times 0.0821 \frac{1\text{atm}}{\text{mole}°\text{K}} \times 323°\text{K}}{\frac{715}{760} \text{ atm} \times 3.41}$$

$= 47.7 \text{ g/mole.}$

Example 19:

One method for the commercial production of chlorine uses the electrolysis of molten sodium chloride. The chemical reaction that occurs is

$$2NaCl\ (1) \rightarrow 2Na(1) + Cl_2(g)$$

Solution:

How many litres of chlorine will be produced from one kg of molten sodium chloride if the gas is measured at 27°C and 1 atm pressure?

From the chemical equation, it is apparent that for each 2 mole (2 × 58.5 = 117.0 g) of NaCl electrolysed, there will be formed 2 moles of Na(I) and one mole (35.5g) of chlorine. This information can be written as

$$2 \text{ moles NaCl} \rightarrow 1 \text{ mole } Cl_2$$

$$117 \text{ g NaCl} \rightarrow \text{mole } Cl_2$$

Therefore,

$$1000 \text{ g NaCl} \rightarrow \frac{1\text{mole}}{117\text{g}} \times 1000 \text{ g } Cl_2$$

$$n = 8.55 \text{ moles } Cl_2$$

Assuming that chlorine obeys ideal gas law,

$$PV = nRT, \text{ we have}$$

$$V = \frac{nRT}{P}$$

$$R = 0.0821 \frac{1 \text{ atm}}{\text{mole } ^\circ K}$$

$$P = 1 \text{ atm}$$

$$T = 273 + 27 = 300^\circ K.$$

Hence, $$V = \frac{8.55 \text{ mole} \times 0.082 \dfrac{1 \text{atm}}{\text{Mole}^\circ K} \times 3000^\circ K}{1 \text{ atm}}$$

$$= 2.106 \times 10^2 1.$$

Example 20:

The density of hydrogen at 0°C and 760 mm Hg pressure is 0.00009 g/cm³. What is the root mean square speed of hydrogen molecules?

Solution:

$$PV = \frac{1}{3} m N \bar{u}^2$$

Therefore, $\bar{u}^2 = 3 \dfrac{PV}{mN}$

As m is the mass of a molecule and N is the number of molecules, therefore, mN, the total mass; so = Mass/Volume = mN/V

and $$\overline{U}^2 = \frac{3P}{\rho}$$

or $$U_{rms} = \left(\overline{U}^2\right)^{1/2} = \left(\frac{3P}{\rho}\right)^{1/2}$$

But $$P = 76 \times 13.59 \times 981 \text{ dynes/cm}^2$$

$$= \left(\frac{3 \times 76 \times 13.50 \times 981}{0.0009 \text{ g/cm}^3} \text{dynes/cm}^2\right)^{1/2}$$

$$= 183.8 \times 10^8 \text{ cm/sec} = 1.838 \text{ km/sec.}$$

Example 21:

A tube 1 metre long and 10.0 cm² cross section is filled with nitrogen at a pressure of .54 atm at 25°. The ends are then sealed with hydrogen permeable material. One end is put in contact with a source of hydrogen at STP while the other end is in contact with vacuum. A steady state is obtained. Estimate the flow rate of hydrogen.

Solution:

$$\frac{1}{A}m_k = -D_k \frac{C_k(2) - C_k(1)}{D}$$

$$\frac{1}{10\,cm^2}m_k = \frac{.67\,cm^2/sec \times .54\ atm}{1\ metre}$$

$$m_k = 0.0217\ y/hour.$$

Example 22:

Calculate the kinetic energy of an Avogadro's number of molecules of gas molecules (that is one mole of gas) at 0°C.

Solution:

Kinetic energy of Avogadro's number of molecules = $N_o\,\overline{KK}$

$$= N_o \cdot \frac{3}{2} RT \left(R = kN_o\right)$$

The value of R is 1.987 cal/mole °K, and T 273°K, thus, kinetic energy of Avogadro's number of molecules

$$= \frac{3}{2}\, 1.987 \frac{cal}{mole.K} \times 273\ K° = 814\ cal/mole.$$

Note that in this calculation, the nature of gas molecules is unimportant. All ideal gases at 0°C will have this amount of kinetic energy.

The average kinetic energy of gas molecule $\left(\overline{KE}\right)$ is generally expressed in ergs and not in calories and at 0°C, it has the value.

$$\overline{KE} = \frac{3}{2}\ KT$$

$$= \frac{3}{2} \times 1.381 \times 10^{-16}\ erg/molecule\ K \times 273°K$$

$$= 5.66 \times 10^{-14}\ erg/molecule.$$

Example 23:

Calculate the root mean square speed of oxygen molecules in the lungs at normal body temperature 37°C.

$$u_{rms} = \left(\frac{3RT}{M}\right)^{1/2}$$

Solution:

In order to give U_{rms} the units of cm/sec, the value of R must be expressed in mechanical energy units, *i.e.*, in ergs (1 erg equal 1 g.cm²/sec²).

$R = 8.31 \times 10^7$ ergs/mole

°K = 8.31×10^7 g.cm² sec^{-2} mole–^{1} °K^{-1}

T = 273 + 37 = 310°K

M = 32 g/mole

Substituting these values in the equation

$$u_{rms} = \left(\frac{3 \times 8.31 \times 10^7 \text{ g.cm}^2 \text{ sec}^{-2} \text{ mole}^{-1} \text{°K} \times 310\text{° K}}{32.0 \text{ g.mole}^{-1}}\right)^{1/2}$$

$= 4.92 \times 10^4$ cm sec^{-1}.

Example 24:

Calculate the pressure exerted by 5.0 moles of carboy dioxide in a 1-litre flask at 47°C, using: (a) the Van der Waals equations, and (b) the ideal gas kw. Compare these results with the experimentally observed pressure of 82 atm.

Solution:

(a) The Van der Waals equation is,

$$\left(P + \frac{n^2a}{V^2}\right)(V - nb) = nRT$$

P = ?

n = 5.0 moles

V = l litre

r = 47 + 273 = 320°K

R = 0.0821 1.atm mole^{-1} °K^{-1}

From the table

a = 3.592 litre² atm. mole^{-2};

b = 0.0427 litre mole^{-2}

Substituting these values

$$\left(P + \frac{25 \times 3.592}{2}\right)(1.0 - 5.0 \times 0.0427) = 5.0 \times 0.0821 \times 320$$

$$(P + 25 \times 3.592)(0.7865) = 5.0 \times 0.0821 \times 320$$

$$P + \frac{5.0 \times 0.0821 \times 320}{0.7865} - 25 \times 3.92 = 77.2 \text{ atm.}$$

(b) For an ideal gas, $PV = nRT$

$$P \times 1.0 = 5.0 \times 0.0821 \times 320$$

$$P = 5.0 \times 0.0821 \times 320 = 131.3 \text{ atm.}$$

The pressure calculated from the Van der Waals equation is closer to the experimental value (85 atm). This is to be expected because the total pressure of the gas is high enough to cause significant deviations from the ideal behaviour.

Example 25:

Two parallel plates are each of 100.0 cm² area and are spaced 1.0 cm apart. One plate is kept at 10.0°C by some means, the other plate at 0°C by being in thermal contact with a mixture of ice and water. The region between the plates is filled with air at atmospheric pressure (K is 0.586×10^{-4} calcm sec. °C). The latent heat of fusion of ice is 79.7 cal/g. Calculate the rate of melting of ice in the freezing mixture.

$$\frac{dQ}{dT} = AK\left(\frac{dT}{dz}\right) = 100 \text{ cm}^2 \times 0.586 \times 10^{-4} \text{ cal/cm sec°C} \times 10°\text{C} \times 60 \text{ sec}$$

$$= \frac{3516 \times 10^{-2}}{797} \text{ g/min} = 0.0441 \text{ g/min.}$$

Example 26:

Argon gas is used in the welding of stainless steel sheets. Calculate its $\left(\overline{C^2}\right)^{1/2}$, $\overline{C}$ and C_p at 25°C.

Solution:

$$\left(\overline{C^2}\right)^{1/2} = \left(\frac{3\,RT}{m}\right)^{1/2}$$

$$= \left(\frac{3 \times 8.314 \times 10^7 \times 298}{40}\right)^{1/2}$$

$$= 4.32 \times 10^4 \text{ cm sec}^{-1}$$

$$\overline{C} = \left(\frac{8\ KT}{\pi m}\right)^{1/2} = \left(\frac{8RT}{\pi m}\right)^{1/2} \left(\frac{8 \times 8.314 \times 10^7 \times 298}{3.14 \times 40}\right)^{1/2}$$

$$= 3.98 \times 10^4 \text{ cm sec}^{-1}$$

$$C_p = \left(\frac{2\ RT}{M}\right)^{1/2} = \left(\frac{2 \times 8.314 \times 10^7 \times 298}{40}\right)^{1/2}$$

$$= 3.52 \times 10^4 \text{ cm sec}^{-1}$$

Example 27:

Calculate the most probable velocity for electron at 25°C.

Solution:

$$C_p = \left(\frac{2 \times 1.38 \times 10^{-16} \times 298}{9.108 \times 10^{-28}}\right)^{1/2}$$

$$= \left(\frac{2 \times 1.38 \times 298}{9.108}\right)^{1/2} \times 10^6 \text{ cm sec}^{-1}$$

$$= 9.504 \times 10^6 \text{ cm sec}^{-1}.$$

Example 28:

Calculate the total number of bimolecular collisions in 1 sec in 1 ml of pure He and pure N_2 at 300°K at 1 atm. The collision diameters of He and N_2 are 2×10^{-8} cm and 4×10^{-8} cm respectively.

Solution:

$$n = \frac{P}{RT} \times 6.02 \times 10^{23} \text{ molecules/ml}$$

$$= \frac{1 \times 6.02 \times 10^{23}}{82.05 \times 300}$$

$$= 2.45 \times 10^{19} \text{ molecules/ml.}$$

$$Z_{12}(\text{He}) = 2 \times (2 \times 10^{-8})^2 (2.45 \times 10^{19})^2 \left(\frac{3.14 \times 8.314 \times 10^7 \times 300}{4}\right)^{1/2}$$

$$= 7 \times 10^{28} \text{ collisions sec}^{-1} \text{ mr}^{-1}$$

$$Z_{12}\,(N_2) = 2(4 \times 10^{-8})^2\,(2.45 \times 10^{-9})^2 \left(\frac{3.14 \times 8.314 \times 10^7 \times 300}{28}\right)^{1/2}$$

$$= 10 \times 10^{28} \text{ collisions sec}^{-1}\text{ mr}^{-1}.$$

Example 29:

The coefficients of viscosity of two liquids at 298 K are 1.408 × 10^{-3} kg m^{-1} s^{-1} and 1.594 × 10^{-3} kg m^{-1} s^{-1} and their densities at the same temperature are 8.07 × 10^{-3} kg m^{-3} and 10.17 × 10^{2}kg m^{-3} respectively. If the time of flow in an Ostwald viscometer for the first liquid is 100 seconds, calculate the time of flow for the second liquid.

Solution:

$$\frac{\eta_1}{\eta_2} = \frac{t_1\rho_1}{t_2\rho_2}$$

or $$t_2\,\frac{\eta_2}{\eta_1} = \frac{t_1\rho_1}{\rho_2}$$

$$= \frac{(1.594 \times 10^{-3}\text{ kg m}^{-1}\text{s}^{-1})\,(100\text{s})\,(8.07 \times 10^2\text{ kg m}^{-3})}{(1.408 \times 10^{-3}\text{ kg m}^{-1}\text{ s}^{-1})\,(10.7 \times 10^2\text{ kg m}^{-3})}$$

Example 30:

A steel ball of density 7.90 gm per cc and 4 mm diameter requires 55 sees to fall a distance of 1 meter through a liquid of density 1.10 gm per cc. Calculate the viscosity of the liquid in poises.

Solution :

According to equation (8), we have

$$\eta = \frac{2r^2\,(\rho - \rho')g}{9v}$$

where, r = Radius = 0.2 cm

v = Velocity = 100/55 cm sec^{-1}

$$\eta = \frac{2 \times (0.2)^2\,(7.90 - 1.10) \times 980}{9 \times \frac{100}{55}} = 32.58 \text{ poises}$$

Example 31:

A sphere of 5 × 10^{-2} cm and density 1.10 gm per cc falls at constant velocity through a liquid of density 1.00 gm per cc and viscosity 1.00 poise. What is the velocity of the falling sphere?

Solution:

Rearranging equation (8) we have

$$v = \frac{2r^2\ (\rho - \rho')g}{9\eta}$$

On substituting

$$v = \frac{2 \times (5 \times 10^{-2})^2 \times (1.10 - 1.00) \times 980}{9 \times 1.00}$$

$$= 5.434 \times 10^{-3} \text{ cm/sec.}$$

Example 32:

Two immiscible liquids, water and chlorobenzene are boiled at 97.89 kNm^{-2} pressure and the mixture boils at 363K. The ratio of the weight of chlorobenzene to water collected in the distillate is 2.47. Calculate the molecular weight of chlorobenzene if vapour pressure of water at this temperature is 70.11 kNm^{-2}.

Solution:

Vapour pressure of chlorobenzene

$$= 97.89 \text{ kNm}^{-2} - 70.11 \text{ kNm}^{-2} = 27.78 \text{ kNm}^{-2}$$

$$\frac{\text{Weight of chlorobenzene}}{\text{Weight of water}} = \frac{P°_A\ M_A}{P°_B\ M_B}$$

i.e.

$$2.47 = \frac{27.78 \text{ kNm}^{-2} \times M_A}{70.11 \text{ kNm}^{-2} \times 0.018 \text{ kgmol}^{-1}}$$

∴ M_A *i.e.*, molecular weight of chlorobenzene

$$= \frac{2.47 \times 70.11 \text{ kNm}^{-2} \times 0.018 \text{ kgmol}^{-1}}{27.78 \text{ kNm}^{-2}} = 0.1122 \text{ kgmol}^{-1}$$

$$\text{Relative molar mass} = \frac{0.1122 \text{ kgmol}^{-1}}{0.001 \text{ kgmol}^{-1}} = 112.2.$$

Example 33:

The radius of a given capillary is 1.05 × 10^{-4} m. A liquid whose density is 0.80g/cm^3 rises in the capillary up to a height of 6.25 × 10^{-2}m. Calculate the surface tension of the liquid assuming the contact angle θ to be zero.

Solution:

Here $r = 1.05 \times 10^{-4}$m $d = 0.80$g/cm^3

$h = 6.25 \times 10^{-2}$m $= 8.0 \times 10^2$kg/m^3

$g = 9.8$ms^{-2}

Since, $\gamma = 1/2$ rhdg

$$\therefore \gamma = \frac{1}{2}(1.05\times10^{-4}\text{m})\ (6.25\times10^{-2}\text{m})\ (8.0\times10^{2}\text{kg m}^{-3})(9.8\text{ms}^{-2})$$

$= 2.5725 \times 10^{-2}$kg s^{-2}

$= 2.5725 \times 10^{-2}$Nm^{-1} (1 N = 1kg ms^{-2})

Example 34:

The surface tension of toluene at 293K is 0.0284 Nm^{-1} and its density at this temperature is 0.866g cm^{-3}. What is the largest radius of the capillary that will permit the liquid to rise 2 × 10^{-2} m ?

Solution:

Since $\gamma = \frac{1}{2}$ rhdg

Therefore $r = \frac{2r}{\text{hdg}}$

Here $\gamma = 2 \times 10^{-2}$ m $h = 0.866$ g cm^{-3}

$\gamma = ?$ $= 8.66 \times 10^2$ kg m^{-3}

$\gamma = 0.0284$ N m^{-1} $g = 9.8$ m s^{-2}

Substituting these values in the above equation, we get

$$r = \frac{2(0.084\ \text{N m}^{-1})}{(2\times10^{-2}\ \text{m})\ (8.66\times1^{2}\ \text{kg m}^{-3})\ (9.8\ \text{ms}^{-2})}$$

$= 3.347 \times 10^{-4}$ kg^{-1} s^{2}m^2

$= 3.347 \times 10^{-4}$m (1N = 1 kg ms^{-2})

$= 0.03347$ cm

Example 35:

At a pressure of 101.3kNm^{-2} a mixture of water and nitrobenzene distils over at 372K. The vapour pressure of water at 372K is 97.70 kN m^{-2}. Estimate the proportion by weight of nitrobenzene in the distillate.

Solution:

$$\frac{\text{Weight of nitrobenzene } W_A}{\text{Weight of water } W_B} = \frac{P^\circ_A\ M_A}{P^\circ_B\ M_B}$$

Molecular weight of nitrobenzene = 0.123 $kgmol^{-1}$

Molecular weight of water = 0.018kg mol^{-1}

$P^\circ_B = P - P^\circ_A = 101.3\ k\ Nm^{-2} - 97.70\ kNm^{-2} = 3.60\ kNm^{-2}$

$$\therefore \quad \frac{W_A}{W_B} = \frac{3.60\ kNm^{-2}}{97.70\ kNm^{-2}} \times \frac{0.123\ gmol^{-1}}{0.018\ kgmol^{-1}}$$

$\therefore$ WA : WB = 0.25 : 1.00

$\therefore$ Proportion of nitrobenzene in the distillate

$$= \frac{0.25}{1.25} \times 100 = 20\%.$$

EXERCISES

1. How is matter classified on the basis of physical characteristics?
2. In which property liquids resemble solids?
3. Name one each of the following.
   (a) solid element
   (b) liquid element
   (c) gaseous element.
4. Name three the most abundant elements on the earth.
5. Write two main characteristics of a chemical compound.
6. Classify the following as pure substances or mixtures.
   (a) Brass (b) Petrol
   (c) Tap water (d) Smoke
   (e) Milk (f) 22 carat gold
7. What kind of impurities can be removed by sedimentation and decantation.
8. Draw a labelled diagram of the experimental set up for sublimation.
9. How would you separate kerosene from a mixture of kerosene and water?

10. Name one typical example of the commercial application of fractional distillation.

11. Mention two applications of the solvent extraction method.

12. Name the stationary phase and the mobile phase in column chromatography.

13. How will you recover nitre from a soil sample rich in nitre?

14. A pure white substance on strong heating gives a colourless gas and a white residue. Is the given substance an element, a compound or a mixture?

15. Give one reason to show that the air is a mixture, and not a chemical compound.

16. Write two main characteristics of a chemical compound.

17. How will you separate the constituents of a mixture containing iodine, sand and sodium chloride?

18. How will you separate the constituents of a mixture containing an oil, alcohol and water?

19. Name the process you would use to separate ammonium chloride from a mixture of ammonium chloride and sodium chloride.

20. What kind of substance be separated from a mixture containing two solids using a magnet?

21. Classify the following statements as the examples of physical properties or chemical properties:

    (a) gold does not get tarnished

    (b) copper is easily drawn into thin wires.

    (c) gallium melts when held in hand.

3

LAWS OF CHEMICAL COMBINATION

INTRODUCTION

During the seventeenth century, the scientists observed that there existed certain regularities and definite pattern in the mass-volume, mass-mass, and volume-volume relationships between the reactants and products.

These similarities were grouped in the form of five laws. These laws deal with the composition of substances by mass or by volume, and are called the laws of chemical combination or laws of stoichiometry. These are described below.

LAW OF CONSERVATION OF MASS

This law was put forward by Lomonosov in 1736. It states that, "*the total mass of matter in any chemical or physical change remains constant though the matter may change its form.*"

In other words, the law may also be stated as,"

"in any chemical reaction, the total mass of the system before and after the reaction is the same. It may however undergo a change in its physical form."

Thus, if in the reaction, $A + B \rightarrow C + D$,

a g of A and b g of B react to give c g of C and d g of D, then

Total mass of reactants = Total mass of products

or $(a + b) = (c + d)$

Another simple statement of the law of conservation of mass is, *"matter is neither created nor destroyed as a result of any chemical or physical change."*

Experimental Verification of the Law of Conversation of Mass

Landolt verified the law of conservation of mass by performing a simple experiment described below.

In this experiment, a H-shaped glass apparatus was used. In one of its limbs, a solution of sodium chloride (NaCl) was filled, while in the second limb a solution of silver nitrate ($AgNO_3$) was filled as shown in Fig. 1.16. The tube was then sealed and weighed precisely. The two reactants were made to react by inverting the tube. The reaction,

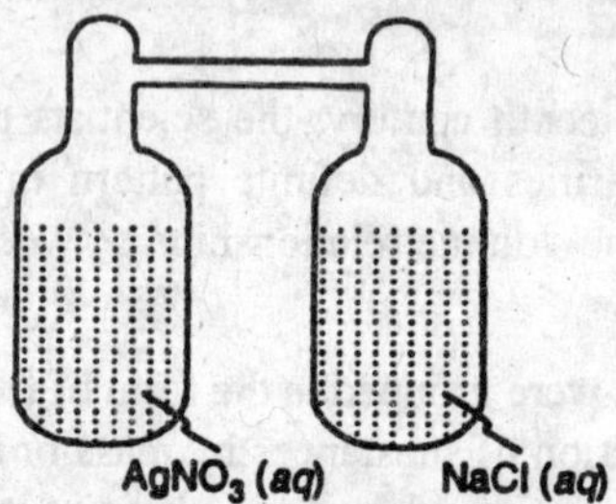

Fig. 3.1 : Landolt's experiment for the verification of the law of conservation of mass.

$$AgNO_3\,(aq) + NaCl(aq) \quad \rightarrow \quad \underset{\text{white ppt.}}{AgCl(s)} + NaNO_3(aq)$$

was allowed to take place. The whole tube was kept for sometime so that the reaction is complete, and the contents of the tube were back to room temperature. The tube was weighed again. From the weights so obtained, it was found that

Weight before the reaction = Weight after the reaction

This verified the law of conservation of mass.

The law of conservation of mass can be utilized in finding out the mass of any reactant or product involved in any reaction as illustrated by the examples given below.

LAW OF DEFINITE PROPORTIONS OR LAW OF CONSTANT COMPOSITION

This law was stated by Louis Proust (1799). In its original form this law states that, *"a chemical compound always consists of the same elements combined together in the same ratio. Irrespective of the method of preparation or the source from where it is taken."*

For example water (H^O) is always found to contain only two elements hydrogen and oxygen combined in a definite mass ratio of 1 : 8 irrespective of its source. Similarly, sodium chloride (NaCl) from whatever source, will always contain 39.32% sodium and 60.68% (by mass) of chlorine.

Carbon dioxide is also found to contain C and is the mass ratio of 12 : 32 (or 3 : 8).

The law of constant proportions can be verified by studying the electrolysis of different samples of water into hydrogen and oxygen. From the experiment, it will be found that the ratio of hydrogen and oxygen is 2 : 1 (by volume) and 1 : 8 (by mass).

It can also be verified by preparing cupric oxide by different methods, and determining the ratio between the masses of copper and oxygen. This is illustrated through the following example.

Conditions under which the law of constant composition does not hold good : The law of constant composition does not hold good under the following conditions.

(a) When a compound is obtained by using different isotopes of the combining elements. For examples, when H_2O formed from ${}_1^2H$ and ${}_8^{16}O$ the mass ratio between hydrogen and oxygen is, 2 : 16 or 1 : 8, whereas when formed using ${}_1^1H$ (deuterium) and ${}_8^{16}O$ the mass ratio between hydrogen and oxygen is 4 : 16 or 1 : 4. Thus, different isotopes of the same element give different mass ratios between the combining elements.

(b) The law of constant composition does not hold good for non-stoichiometric components.

This is because, the non-stoichiometric compounds do not have fixed composition.

LAW OF MULTIPLE PROPORTIONS

This law was enunciated by Dalton in 1803. This law applies to the situations where one element combines with another to give more than one compound of different stoichiometry. For example, copper reacts with oxygen to give cuprous oxide (Cu_2O) and cupric oxide (CuO). Nitrogen reacts with oxygen to give five oxides, *e.g.*, N_2O (nitrous oxide), NO (nitric oxide), N_2O_2 (dinitrogen trioxide). NO_2 (nitrogen dioxide) and N_2O, (nitrogen pentaoxide).

This law states that,

"When two elements combine to form two or more compounds, then the different masses of one element which combine with a fixed mass of the other, bear a simple ratio to one another."

As referred to above, the masses of copper for a certain fixed (say 16 g) of oxygen in the case of the two oxides so formed are,

Cu_2O	CuO
$(2 \times 63.5) : 16$	$63.5 : 16$

Thus, the ratio of masses of copper combined with a fixed amount of oxygen (*e.g.*, 16g) is 2 : 1 (a simple ratio). The law of multiple proportion is thus verified.

For nitrogen and oxygen to give above mentioned five oxides, the ratio can be obtained as follows.

Table 3.1

Oxygen of nitrogen	*Mass (g) of*		*Mass of oxygen for 14 g of nitrogen*	*Ratio of the masses of oxygen*
	Nitrogen	*Oxygen*		
N_2O	28	16	8	1
NO	14	16	16	2
N_2O_3	28	48	24	3
NO_2	14	32	32	4
N_2O_5	28	80	40	5

Thus, a simple ratio of 1 : 2 : 3 : 4 : 5 exists for the masses of oxygen which combine with a fixed (14 g) mass of nitrogen. It must be remembered that the law of multiple proportions hold good only when the same pair of isotopes of the combining elements is used for obtaining different compounds.

LAW OF RECIPROCAL PROPORTIONS OR LAW OF EQUIVALENT PROPORTIONS

This law was put forward by Richter in 1792. It states that,

"If three elements A, B and C mutually combine to give binary compounds AB, BC and CA, then the ratio of the masses of B and C when they combine together, is a simple multiple of the mass ratio of B an dC in which they combine separately with a fixed amount of A."

For example, copper reacts with oxygen and sulphur to give copper oxide and copper sulphide respectively. Sulphur and oxygen also react with each other to give. Then,

In CuS, Cu : S = 63.5 : 32

In CuO, Cu : O = 63.5 : 16

Therefore, S : O = 32 : 16

or S : O = 2 : 1

Now in SO_2, S : O = 32 : 32

or S : O = 1 : 1

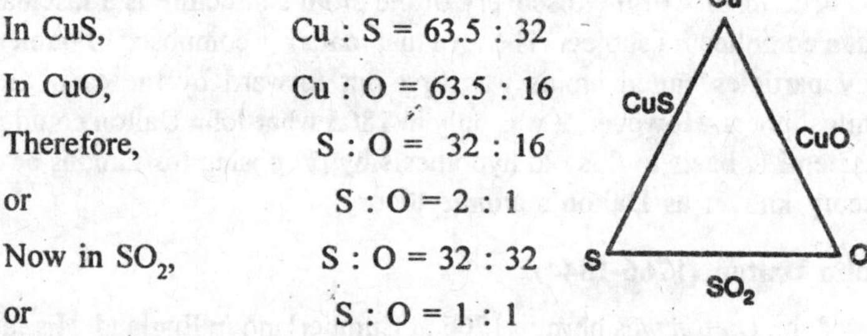

Fig. 3.2

Thus, the ratio between the two ratios is

2/1 : 1/2 = 2 : 1, *i.e.*, a simple multiple ratio.

GAY-LUSSAC'S LAW OF GASEOUS VOLUMES

This law put forward by the French chemist, Gay-Lussac, establishes an interesting volume-volume relationship between the volumes of the reactants and products. According to this law,

"Under constant temperature and pressure, the volume of the gaseous reactants bear a simple ratio between themselves and also with the volumes of the products, if gaseous."

For example, in the reactions

(i)$H_2(g)$ + $Cl_2(g)$ → $2HCl(g)$
1 vol 1 vol 2 vol

The ratio of the volumes of hydrogen, chlorine, and hydrogen chloride is 1 : 1 : 2 (a simple ratio).

(i)$N_2(g)$ + $3H_2(g)$ → $2NH_3(g)$
1 vol 3 vol 2 vol

The volume ratio of 1 : 3 : 2 is a simple ratio.

(iii) $2H_2(g) + O_2(g) \rightarrow 2H_2O(g)$

2 vol 1 vol 2 vol

The volume ratio of 2 : 1 : 2 is a simple ratio.

Thus, it appears that the law of gaseous volumes (or law of combining volumes) is actually the law of definite proportions by volume.

DALTON'S ATOMIC THEORY

The history of the discovery of the atom's structure is a fascinating but a complicated subject. The idea that matter is composed of numerous tiny particles called atoms was first put forward by the early Greek philosophers. However, it was only in 1805 what John Dalton could give a scientific basis to this old hypothesis by proposing his famous atomic theory known as Dalton's atomic theory.

John Dalton (1766-1844)

John Dalton was born in 1766 at Cumberland in England. His father was a poor weaver. John after his schooling, began teaching at the age of 12. He became school principal at 18. Later, he taught in Colleges at Glasgow, Edinburgh, London and Manchester. His atomic theory is recognised even today as one of the most brilliant work in the field of science. Dalton, for the first time suggested a shorthand method for denoting both elements and compounds.

Postulates of the Dalton's Atomic Theory

The main postulates of the Dalton's atomic theory are,

(i) All substances are made up of tiny, indivisible particles, called atoms. The word atom was derived from the Greek word atomos (meaning - indivisible).

(ii) Atoms cannot be created, divided or destroyed during any chemical or physical change (the law of conservation of mass).

(iii) Each element is composed of its own kind of atoms.

(iv) The atoms of a given element are alike, and have the same mass. The atoms of different elements differ in mass and properties.

(v) The atoms combine with each other in simple whole-number ratios to form compound atoms.

Need to Modify the Dalton's Theory

Until the end of 19th century, Dalton's atomic model was widely accepted, because it explained many of the chemical and physical laws known at that time. Around the turn of the century, however, certain experimental facts were discovered which went against the Dalton's theory. Some of the facts which went against the Dalton's theory were

(i) The discovery of radioactivity showed that small charged particles are emitted from a radioactive element. This means, that the atom is not the ultimate particle.

(ii) The discharge-tube experiments, based on the passage of electricity through gases under reduced pressure showed the existence of the charged particles, viz; electrons and protons in an atom. This indicates that the atom can be split further into smaller particles.

(iii) Discovery of isotopes indicated that the atoms of the same element are not identical in all respects.

(iv) The discovery of the nuclear reactions has made it possible to convert matter into energy. Nuclear fission and fusion have shown that the atoms can be created or destroyed.

In the light of these facts, it became necessary to modify some of the statements of the Dalton's theory, while retaining the basic framework of the theory. Accordingly, it can be said that

(a) Although an atom can be split into particles such as electrons, protons and neutrons, but atom is the smallest unit of matter which takes part in all chemical processes.

(b) The smallest particle of any compound which shows its characteristic properties and is capable of independent existence is a molecule, Dalton's compound atom. However, atoms of the noble gases can exist independently.

DALTON'S ATOMIC THEORY AND THE LAWS OF CHEMICAL COMBINATION

Dalton's theory could explain successfully the laws of conservation of mass and the law of constant composition. Incidentally, only these two laws were known at that time. The theory could predict the law of multiple proportions.

Dalton's Theory and the Law of Conservation of Mass

Dalton's theory postulates that atoms are indestructible during a chemical or physical process. So, during chemical reactions, no atom is created or destroyed. As a result, the total mass of all the atoms before and after the reaction remains the same. Thus, the Dalton's theory explains the law of conservations of mass.

Dalton's Theory and the Law of Constant Composition

According to the Dalton's theory, a compound is formed when a certain fixed number of atoms of each combining element combine with each other. The theory also postulates that all atoms of a given element have the same fixed mass. So, in any compound, each element should have a fixed mass (no. of atoms x mass of an atom: both being fixed). Thus, a compound always contains the same elements combined in the same fixed proportion by mass.

Let us suppose, a compound $A_n B_m$ is formed when n atom of A and m atoms of B combine together: n and m are the whole numbers. If a is the mass of one atom of A, and b is the mass of one atom of B, then

Total mass of A in the compound $A_n B_m = n \times a = na$

Total mass of B in the compound $A_n B_m = m \times b = mb$

and, Mass of the compound $A_n B_m = (na + mb)$

Therefore,

$$\text{Percentage of A in the compound} = \frac{na \times 100}{(na + mb)}$$

$$\text{Percentage of B in the compound} = \frac{na \times 100}{(na + mb)}$$

Since na and mb are fixed in the case of any compound, hence the compound $A_n B_m$ contains a fixed mass of A and B, or the percentages of A and B in the compound $A_n B_m$ are fixed. This is what the law of constant composition says.

Dalton's Theory and the Law of Multiple Proportions

According to the Dalton's theory, atoms combine with each other in a simple whole-number ratios. Also, all atoms of a given element have the same fixed mass. So, when the atoms of two elements say A and B, combine to give more than one compound, they would do so in a simple whole number ratio, *i.e.*, 1 : 1, 1 : 2, or 2 : 1 etc. Let us suppose, they do so in the ratio 1 : 1 and 1 : 2, *i.e.*, A and B combine to form two compounds, AB and AB_2.

Let a and b be the mass of each atom of A and B respectively. Then,

Compound	*Mass of A*	*Mass of B*	
AB	a	b	
AB_2	a	2b	Ratio between the masses of B = b : 2b = 1 : 2

Thus, the ratio between the masses of B which combine with a fixed mass of A is 1 : 2. This is a simple ratio as postulated by the law of multiple proportions.

AVOGADRO'S LAW

An Italian physicist, Amedeo Avogadro in 1811 suggested that the smallest particle in gases were not atoms but molecules. These molecules could contain two, three or more atoms. The molecules of the noble gases however, contain only one atom.

While studying the physical properties of gases, Avogadro proposed his famous hypothesis, known as Avogadro's hypothesis. This hypothesis (now known as Avogadro's law) states that,

"Under similar conditions of temperature and pressures equal volumes of all gases contain the same number of molecules."

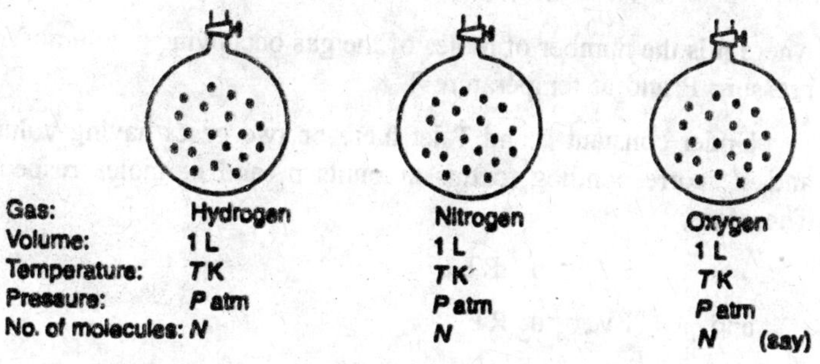

Fig. 3.3

For example, if equal volumes of different gases are held in separate containers under similar conditions of temperature and pressure, then each container will contain the same number of molecules. *The total mass of each gas in the container may differ from each other.*

It has been found experimentally that one mole of all gases under similar conditions of temperature and pressure occupy the same volume. This volume is called molar volume (V_m). Molar volumes of all gases under normal pressure (1 atm) and normal temperature (273K) have a value of 22.4 litre per mole.

AMEDEO AVOGADRO (1756-1856)

Amedeo Avogadro born in Turin (Italy) obtained degrees in Philosophy at the age of 13, in Jurisprudence at 16 and in Church law at 20. In 1809, Avogadro became professor of Physics at the Royal College at Vercelli. In 1811, he postulated his famous hypothesis which is now known as Avogadro's law. In 1820, Avogadro was appointed professor of Mathematical Physics at the University of Turin.

It has been experimentally found that the molar volume of each gas (V_m) contains 6.023 × 1023 molecules. This number is called the Avogadro's number, and is denoted by N_A. Avogadro's number (N_A) is also called Avogadro's constant.

Since the later developments have fully confirmed the Avogadro's hypothesis, hence it in practice, enjoys the status of a law.

Derivation of the Avogadro's Law from Gas Equation

The gas equation is

$$PV = nRT$$

where n is the number of moles of the gas occupying a volume V under pressure P and at temperature T K.

Under constant P and T let there be two gases having volume V_1 and V_2 corresponding to this amounts n_1 and n_2 moles respectively. Therefore,

$$PV_1 = n_1, RT$$

and $$PV_2 = n_2 RT$$

If N_1 and N_2 denote the number of molecules of gas 1 and 2 respectively then,

$$n_1 = \frac{N_1}{N_A}$$

$$n_2 = \frac{N_2}{N_A}$$

where N_A is the Avogadro's number. Substitution for n_1 and n_2 in the above equations gives,

$$PV_1 = \frac{N_1}{N_A}.RT$$

$$PV_2 = \frac{N_2}{N_A}.RT$$

Dividing first equation with-the, second,

$$\frac{V_1}{V_2} = \frac{N_1}{N_2}$$

If, $V_1 = V_2$

Then, $N_1 = N_2$

Thus, the number of molecules of both the gases must be equal. It can be generalized, *i.e.*, under similar conditions of temperature and pressure equal volumes of all gases contain equal number of molecules. This is exactly what the Avogadro's law postulates.

APPLICATIONS OF AVOGADRO'S LAW

Avogadro's law has been found very useful in chemistry. The usefulness of the Avogadro's law lies in the fact that it differentiates between an atom and a molecule. Some important applications of the Avogadro's law are described below.

Derivation of the Gay-Lussac's Law of Gaseous Volume

Consider the reaction,

$$a\,A + b\,B \rightarrow c\,C + d\,D$$

Let, a molecules of A react with b molecules of B, and there be n molecules in a unit volume of A. Under similar conditions of temperature and pressure, B should also have n molecules in one unit of volume (the Avogadro's hypothesis). Then,

Volume of A involved in the reaction = $1/n \times a$ units

and, Volume of B involved in the reaction = $1/n \times b$ units

Therefore, the ratio of volumes of the reacting gases A and B is,

$$\frac{a}{b} : \frac{b}{n} \text{ or } a : b$$

Since a and b are whole numbers (molecules cannot be in fractions), hence the volumes of the reacting gases bear a simple whole number ratio to each other. This is the Gay-Lussac's law of gaseous volumes.

Deduction of the Atomicity of the Elementary Gases

The number of atoms present in a molecule of a gas is called its atomicity. For example, hydrogen, oxygen, nitrogen gases exist as diatomic molecule, *i.e.*, a molecule of hydrogen, oxygen and nitrogen, each contains two atoms.

To illustrate the determination of the atomicity of an elementary gas, we consider the reaction between hydrogen and chlorine to form hydrogen chloride.

It has been observed experimentally that the hydrogen and chlorine react in equal volumes.

Then, one can write,

Hydrogen	+	Chlorine	→	Hydrogen chloride	
1 vol		1 vol		2 vol	
n molecules		n molecules		2n molecules	(Avogadro's
1/2 molecule		1/2 molecule		1 molecule	hypothesis)

Since, one molecule of hydrogen chloride contain one atom of hydrogen and one atom of chlorine, hence,

1/2 molecule of hydrogen = 1 atom of hydrogen

or, 1 molecule of hydrogen = 2 atoms of hydrogen

Similarly, 1/2 molecule of chlorine = 1 atom of chlorine

or, 1 molecule of chlorine = 2 atoms of chlorine

Thus, the atomicity of both hydrogen and chlorine is two.

Derivation of the Relationship Between Molar Mass and Vapour Density

The vapour density (V.D.) of a gas is defined as the ratio between the mass of a certain volume of the gas, to the mass of the same volume of hydrogen gas under similar conditions of temperature and pressure. Thus,

$$\text{Vapour density (V.D.)} = \frac{\text{Mass of V mL of the substance in the gaseous state}}{\text{Mass of V mL of hydrogen under similar conditions}}$$

According to the Avogadro's hypothesis, *equal volumes of gases under similar conditions should contain equal number of molecules.* Therefore, if V mL of any gas contain n molecules, then

$$\text{V.D.} = \frac{\text{Mass of n molecules of the gaseous substance}}{\text{Mass of n molecules of hydrogen}}$$

$$\text{or V.D.} = \frac{\text{Mass of 1 molecules of the gaseous substance}}{\text{Mass 1/2 of 1 molecules of hydrogen}}$$

Since, one molecule of hydrogen contains 2 atoms, hence

$$\text{V.D.} = \frac{\text{Mass of 1 molecules of the gaseous substance}}{2 \times \text{Mass of 1 atom of hydrogen}}$$

As per definition

$$\frac{\text{Mass of 1 molecules of the gaseous substance}}{\text{Mass of 1 atom of hydrogen}}$$

$= \text{Molar mass of the substance}$

Hence, Molar mass of the substance = 2 × V.D. of the substance

In general, therefore

Molar mass = 2 × Vapour density

Determination of the Molecular Formula of a Gaseous Compound

The Avogadro's hypothesis can also be used for determining the molecular formula of a gaseous compound, if the volumes of the reactants and products are known. This is illustrated by taking example of the reaction, between nitrogen and hydrogen gases to form ammonia. Experimentally, it is found that one volume of nitrogen reacts with three volumes of hydrogen to form two volumes of ammonia. Then, one can write,

	Nitrogen	+	Hydrogen	→	Ammonia	
	1 vol		3 vol		2 vol	
	n molecules		3n molecules		In molecules	(Avogadro's
or	1/2 molecule		3/2 molecule		1 molecule	hypothesis)
	1 atom		3 atom		1 molecule	

Because nitrogen and hydrogen gases are diatomic.

Thus, one molecule of ammonia contains one atom of nitrogen and three atoms of hydrogen. Therefore, the molecular formula of ammonia is NIL.

ATOMIC AND MOLECULAR MASSES

Despite the confirmed existence of an atom, it cannot be seen or isolated. Therefore, a single atom cannot be weighed. Actual mass of an atom is extremely small. For example, one atom of hydrogen has a mass of 1.673 × 10"24 g. It is not possible to measure such small masses. Avogadro in 1811 suggested that the atomic and molecular masses can be expressed on an atomic mass scale based on the comparison of atomic/molecular masses with the mass of a reference atom. This method for expressing atomic/molecular masses is described below.

ATOMIC MASS SCALE AND ATOMIC MASS UNIT

In this method, the atomic masses are expressed relative to that of a standard reference atom. Earlier, hydrogen and oxygen atoms were used as the reference atoms for describing the atomic masses. In 1957, Olander and Nier proposed an atomic mass scale based on the choice of carbon-12 isotope, (designated as ${}^{12}_{6}C$ or simply as ${}^{12}C$).

IUPAC accepted this scale to be used for expressing atomic and molecular masses in 1961. In this scale of atomic masses, the ${}^{12}C$ isotope has been assigned an atomic mass of 12.000000 atomic mass units. Since, ${}^{12}C$ isotope has been assigned an atomic mass of 12 units exactly, the atomic mass unit may be defined as follows:

"The mass equal to 1/12 th of the mass of a "C atom is called one atomic mass unit" The atomic mass unit is abbreviated as amu, and denoted by the symbol u.

$$1 \text{ atomic mass unit} = 1 \text{ u} = \frac{\text{Mass of a } ^{12}\text{C atom}}{12}$$

Absolute mass of a ^{12}C atom is 1.9924×10^{-23}g. So,

$$1 \text{ u} = \frac{1.9924 \times 10^{23}}{12} = 1.66 \times 10^{-24}\text{g} = 1.66 \times 10^{-27}\text{kg}$$

As we have seen above, the atomic and molecular masses are expressed on a relative scale based on the mass of a ^{12}C atom. Thus, the atomic and molecular masses are relative masses. Before we discuss the formal definitions of atomic and molecular masses, let us know that the atomic mass of an element is not the mass of one atom of that element. Likewise, the molecular mass of a compound is not equal to the mass of one molecule of that compound. As we know, most of the naturally-occuring elements occur as mixtures of their various isotopes. Isotopes of an element are different atoms of the same element which have different atomic masses. So, in any sample of an element, there might be atoms of different masses existing together. *The mass of which of the atoms (isotopes) should be taken as the atomic mass of that element?* Obviously, the only solution is to take the weighted-average of the mass of these isotopes. Thus, all atomic and molecular masses in fact are the average relative masses.

For example, oxygen occurs in nature as a mixture of three isotopes having atomic masses of 15.995 u, 16.999 u and 17.999 u. These isotopes are designated as ^{16}O, ^{17}O and ^{18}O respectively. The relative proportions of these isotopes in the mixture are, 99.763%, 0.037% and 0.200% respectively. Then, the atomic mass of oxygen (as it occurs in nature) is given by the weighted average of the atomic masses of the three isotopes. Thus,

Atomic mass of oxygen

$$= \frac{(99.763 \times 15.995) + (0.037 \times 16.999) + (0.20 \times 17.9990\text{u}}{100} = 15.999\text{u}$$

IUPAC RECOMMENDATIONS REGARDING ATOMIC AND MOLECULAR MASSES

According to IUPAC, the atomic and molecular masses are defined as follows:

Relative Atomic Mass (A_r)

The relative atomic mass (A_r) of an element is defined as the average relative mass of an atom of the element compared with an atom of ^{12}C taken as 12 atomic mass units. Thus,

Relative atomic mass of an element (A_r)

$$= \frac{\text{Average mass of 1 atom of the element}}{1/12 \times (\text{Mass of one } ^{12}\text{C atom})}$$

The relative atomic mass is denoted by A_r. The relative atomic mass is a pure number, and hence it has no unit.

Table 3.2 : Relative Atomic Masses of Some Common Elements (base ^{12}C = 12.000 u)

Element		*Relative atomic mass (A_r)*		*Element*		*Relative atomic mass (A_r)*	
Name	*Symbol*	*Exact*	*Common value**	*Name*	*Symbol*	*Exact*	*Common value**
Hydrogen	H	1.008	1.0	Sulphur	S	32.06	32.0
Carbon	C	12.010	12.0	Sodium	Na	22.99	23.0
Oxygen	O	15.999	16.0	Silver	Ag	107.87	108.0
Nitrogen	N	14.007	14.0	Copper	Cu	63.54	63.5
Chlorine	Cl	35.45	35.5				

* these values are for use for solving the numerical problems.

The relative atomic mass of an element indicates the number of times one atom of that element is heavier than 1/2th of a ^{12}C atom. For example, the average relative mass of chlorine is 35.5. This means that an atom of chlorine on average is 35.5/12 times heavier. Than one atom of ^{12}C.

Names, symbols and relative atomic masses of some common elements are presented in Table 3.2.

A complete list of elements, and their relative atomic masses is presented on the inside cover of the book.

Atomic Mass (A)

The average mass of an atom of an element in atomic mass units is known as its atomic mass.

Atomic mass of an element is designated as A. Since, atomic mass is actual mass, it may have the unit of mass, viz., kg, g or u. However, as per convention, the atomic masses are expressed in the atomic mass units.

From Eq. 3 one can write,

Average mass of 1 atom of an element = Relative atomic mass ×

$$\left(\frac{\text{Mass of one } ^{12}\text{C atom}}{12}\right)$$

Thus, Atomic mass (A) of an element = $A_r \times 1\ u = A_r\ u$

Thus, we see, that the magnitudes of the relative atomic mass, and the atomic mass are equal. The two differ only in their units.

Relative Molecular Mass

Molecular masses are very small. These also cannot be measured directly. Therefore, the molecular masses are also expressed as the relative molecular masses. The relative molecular mass (M_r) of any substance is defined as follows.

"The relative molecular mass of a substance is the average relative mass of its molecule as compared with the mass of a ^{12}C atom taken as 12 units."

Relative molecular mass,

$$M_r = \frac{\text{Average mass of a molecule of the substance}}{1/12 \times (\text{Mass of an atom of } ^{12}\text{C}}$$

The relative molecular mass (A_2.) is a number and has no unit.

Molecular Mass

The molecular mass of a substance may be defined as follows.

"The average mass of a molecule of a substance in atomic mass unit is called its molecular mass (M)."

The molecular mass is actually mass. So it can have the units of mass, that is kg, g or atomic mass unit. However, as per convention, the molecular masses are described hi atomic mass unit.

From the above relationship, one can write

Average mass of a molecule of a substance

$$= \text{Relative molecular mass} \times \frac{\text{Mass of an atom of } ^{12}\text{C}}{12}$$

or Molecular mass, $(M) = M_r \times 1\ u = M_r\ u$

It may be noted, that the magnitudes of the relative molecular mass and molecular mass are equal. The two differ only in their units.

Molar Mass

The average mass of one mole of any material is called its molar mass. Mathematically,

$$\text{Molar mass (M)} = \frac{\text{Mass of the substance}}{\text{Amoun of the substance (in moles)}} = \frac{m}{n}$$

Since, the unit of mass is kg or g, and that of n is mol, hence the unit of molar mass is gram per mole (g/mol), or kilogram per mole (kg/mol).

One mole of any material contains 6.023×10^{23} chemical units. The chemical units may be atoms, ions or molecules. Thus,

Molar mass of any substance = Mass of 6.023×10^{23} chemical units of that substance

For example,

Molar mass of hydrogen gas (H_2) = Mass of 6.023×10^{23} molecules of hydrogen (H_2) and

Molar mass of hydrogen atom (H) = Mass of 6.023×10^{23} atoms of hydrogen (H)

Gram-Atomic Mass and Gram-Molecular Mass

Historically, the terms gram-atomic mass (or gram-atom), gram-molecular mass (or gram-mole) and gram-ionic mass (or gram-ion), were used as the practical units of atomic and molecular masses. These terms are no longer used. The practical unit presently used for describing the atomic and molecular masses is mole. So, although the terms, gram-atomic mass and gram molecular mass are not relevant now, but wc describe these just because of their historical significance.

Gram-Atomic Mass

The number of grams of an element numerically equal to its relative atomic mass (or

atomic mass) is termed as the gram-atomic mass of that element. For example, the relative atomic mass of hydrogen is 1.008. So, the gram-atomic mass of hydrogen is 1.008g.

In terms of the mole concept, the gram-atomic mass of an element is equal to the mass (in grams) of 6.02×10^{23} atoms of that element. The number 6.02×10^{23} is termed as the Avogadro's number. So, the gram-atomic mass of an element is equal to the mass (in gram units) of the Avogadro's number of atoms of that element.

Gram-molecular Mass

The number of grams of a substance numerically equal to its relative molecular mass is termed as the gram-molecular mass of that element. For example, the relative molecular mass of hydrogen is 2.016. So, the gram-molecular mass of hydrogen is 2.016g.

In terms of the mole concept, the gram-molecular mass of a substance is equal to the mass (in gram units) of 6.02×10^{23} molecules of that substance. The number 6.02×10^{23} is termed as the Avogadro's number. So, the gram-molecular mass of a substance is equal to the mass (in gram units) of the Avogadro's number of molecules of that substance,

CALCULATION OF MOLECULAR MASS FROM ATOMIC MASSES

The molecular mass of any substance is calculated by adding together the atomic masses of all the atoms present in one molecule of the substance. This is illustrated through the following examples.

(i) *Molecular mass of water* : The molecular formula of water is H_2O. So,

Molecular mass of waters = (2 × Atomic mass of hydrogen) + (1 × Atomic mass of oxygen)

$$= (2 \times 1\ u) + (1 \times 16\ u) = 2\ u + 16\ u = 18\ u$$

(ii) Molecular mass of sulphuric acid (H_2SO_4). The molecular formula of sulphuric acid in H_2SO_4. So,

Molecular mass of sulphuric acid = (2 × Atomic mass of H) + (1 × Atomic mass of sulphur) + (4 Atomic mass of oxygen)

$$M(H_2SO_4) = (2 \times 1\text{ u}) + (1 \times 32\text{ u}) + (4 \times 16\text{ u})$$

$$= 2\text{u} + 32\text{u} + 64\text{u} = 98\text{u}$$

So, the molecular mass of H_2SO_4 is 98 u.

(iii) Molecular mass of sucrose ($C_{12}H_{22}O_{11}$). The molecular formula of sucrose (canesugar) is $C_{12}H_{22}O_{11}$. So,

Molecular mass of sucrose ($C_{12}H_{22}O_{11}$) = (12 × Atomic mass of C) + (22 × Atomic mass of H) + (11 × Atomic mass of O)

$$= (12 \times 12\text{ u}) + (22 \times 1\text{ u}) + (11 \times 16\text{ u})$$

$$= (144\text{ u}) + (22\text{ u}) + (176\text{ u}) = 342\text{ u}$$

THE MOLE CONCEPT

The term mole is derived from the Latin word meaning a pile or mass of stones placed in the sea, often as a breakwater. Likewise *the term mole in chemistry represents a pile or mass of atoms, molecules, ions or electrons*. In this section, we discuss the mole concept and its applications.

We buy certain things either by number or by mass. For example, we generally buy oranges, eggs, bananas etc., by number, viz., by dozen (1 dozen = 12 pieces). We buy most of our vegetables by mass, viz., gram or kilogram. The unit we choose to express the quantity is just a matter of convenience. Incidentally, you will be amazed to know that in supermarkets of USA/Canada, bananas are sold by weight and not by numbers as done here in our country. So, look the choice of the unit depends upon the choice of the people like you and me.

In much the same way we buy a dozen of eggs, or a kilogram of sweets, a chemist deals with a mole of atoms, molecules, ions or electrons. Thus,

"The mole is a concept of quantity in terms of number and mass, which relates the mass of a material to the number of atoms, ions or molecules in it"

The mole, as per definition accepted internationally, is defined as follows.

"The mole is the amount of a substance which contains the same number of chemical units (atoms, molecules or ions) as there are atoms in exactly 12 grams of pure carbon-12."

12 g of carbon-12 is found to contain 6.023×10^{23} atoms of carbon-12. Thus, a mole represents a collection of 6.023×10^{23} chemical units (atoms, molecules or ions). The number 6.023×10^{23} is called the Avogadro's number. The Avogadro's number is denoted by N_A or L. Most commonly the symbol N_A is used.

How long is the number 6.023×10^{23}? If you spend Rs. 10 lakh per second, to spend Rs. 6.023×10^{23} (one mole rupees) you will need 19 billion years ! How many generations could survive?

Thus, a mole represents the quantity of material which contains one Avogadro's number (6.023×10^{23}) of chemical units (atoms, molecules, or ions) of any substance.

It is important to note that while using the unit mole, it is necessary to specify the chemical unit also. For example,

One mole of hydrogen atoms

$= 6.023 \times 10^{23}$ atoms of hydrogen

One mole of hydrogen molecules

$= 6.023 \times 10^{23}$ molecules of hydrogen

One mole of carbon dioxide

$= 6.023 \times 10^{23}$ molecules of carbon dioxide

One mole of electrons

$= 6.023 \times 10^{23}$ electrons

One mole of sodium ions (Na^+)

$= 6.023 \times 10^{23}$ Na^+ ions

Symbol of the mole unit : Each unit has been given a standard symbol. For example, gram is denoted by g, a kilogram by kg, a metre by m, and newton by N (newton is the unit of force). The unit of mole is given a symbol mol. So, if you want to express one mole, you may write it as 1 mol.

Please note, that in the text, one uses the word mole, but when used as a unit, then it is denoted as mol.

Mole Concept for Ionic Substances

In ionic compounds there is no discrete molecule. So, it is meaningless to talk of a molecule or molecular formula of such compounds. An ionic compound on the other hand is described by a simple formula commonly called as *empirical formula or stoichiometric formula.*

"Empirical formula (or stoichiometric formula) of a substance is the simplest formula which gives the lowest whole-number ratio between the number of atoms of different elements present in the substance."

For example, we know that sodium chloride is a giant-structure containing sodium (Na^+) and chloride (Cl^-) ions. Thus, it may be represented by a formula $(Na^+Cl^-)_n$ where n is a large number. The actual number of Na^+ ions and Cl^- ions in a sample of sodium chloride depends upon the size of the sample. But, in all samples, small or big, the ratio between the number of Na^+ ions and Cl^- ions remains 1 : 1. So, the simplest formula for sodium chloride is Na^+Cl^- or only as NaCl. Thus, the *empirical formula (or stoichiometric formula) of sodium chloride is NaCl.* So, when we use mole concept for ionic compounds, we should know that

One mole of an ionic compound = 6.023 x 1023 formula units of the compound

Thus, for sodium chloride and barium chloride we can write,

One mole of sodium chloride = 6.023×10^{23} Na^+Cl^- units

= 6.023×10^{23} Na^+ ions + 6.023×10^{23} Cl^- ions

and

One mole of barium chloride = 6.023×10^{23} $BaCl_2$ units

= 6.023×10^{23} Ba^{2+} ions + 2 × 6.023 × 1023 Cl^- ions

Molar mass of ionic substances : In the case of ionic compounds, the molar mass in grams is numerically equal to its formula mass.

One mole of sodium chlorides = Mass of 6.023×10^{23} Na+Cl^- ion pairs

= (Mass of 6.023×10^{23} Na^+ ions) + (Mass of 6.023×10^{23} Cl^- ions)

= (Molar mass of Na atoms) + (Molar mass of Cl atoms)

= (23 g/mol) + (35.5 g/mol) = 58.5 g/mol

Molar Volume

The volume occupied by one mole of any substance is called its molar volume. Molar volume is denoted by V_m. Molar volume of a substance depends upon temperature and pressure. Molar volume of gaseous substances change appreciably with temperature and pressure. The molar volume of all gaseous substance at 273 K and under 1 atm pressure (NTP conditions) is found to be 22.4 litre or 22400 mL. Thus, one mole of any gaseous substance at 273 K and 1 atm pressure occupies a volume equal to 22.4 L or 22400 mL. The unit of molar volume is litre per mol (L/mol) or millilitre per mol (mL/mol).

Calculation of Moles In Certain Volume of a Gas

To calculate the number of moles of a gaseous substance in a certain volume, we use the relationship, that one mole of any gaseous substance under NTP conditions occupies 22.4 litre (L) or 22400 mL of volume. Let,

Volume of any gaseous substance at NTP = V L

Molar volume of a gas under NTP = 22.4 L/mol

So,

$$\text{No. of moles of the substance in V L} = \frac{\text{Volume (in L) of the gaseous substance at NTP}}{22.4 \text{ L / mol}}$$

$$= \frac{V}{22.4} \text{ mol}$$

and

$$\text{No. of molecules of the substance in V L} = \frac{V}{22.4} \text{ mol} \times 6.023 \times 10^{23} \text{ molecule/mol}$$

$$= \frac{V}{22.4} \times 6.023 \times 10^{23} \text{ molecule}$$

CALCULATION OF MOLES IN CERTAIN MASS OF A SUBSTANCE

Number of moles of a substance in a certain mass of that substance may be obtained as follows.

Let, Mass of the substance = g

Molar mass of the substance = M g/mol

Then, No. of moles of substance = $\frac{W\ g}{M\ g/mol} = \frac{W}{M}$ mol

or No. of moles of the substance

$$= \frac{\text{Mass of the substance in g}}{\text{Molar mass of the substance in g / mol}}$$

The number of molecules in certain mass (say, W g) of a substance having molar mass of M g/mol is given by the following relationship.

No. of molecules in a sample

$$= \frac{\text{Mass of the substance}}{\text{Molar mass of the substance}} \times \text{Avogadro's number}$$

$$= \frac{W}{M} \times 6.023 \times 10^{23}$$

CHEMICAL FORMULAE (EMPIRICAL AND MOLECULAR FORMULAE)

Each chemical compound is known by a specific name. While describing chemistry, such names are repeated quite often. Writing full name of a compound repeatedly is time consuming and inconvenient. Also, the name of a compound does not provide any information about the chemical composition of the compound. To meet these requirements, each substance is denoted by its chemical formula. Therefore, a chemical formula is a shorthand notation of a compound.

Certain elements and covalent compounds which exist as discrete molecules, can be described by their molecular formulae.

In ionic compounds (such as NaCl) and network covalent compounds (*e.g.*, diamond) there is no discrete molecule. Therefore, ionic and network covalent compounds are described by their stoichiometric formulae. Stoichiometric formula is also called empirical formula.

There are three types of chemical formulae.

(a) Molecular formula

(b) Empirical formula

(c) Structural formula

Only the first two types of chemical formulae are described here.

MOLECULAR FORMULA

The smallest unit of a substance capable of independent existence and showing all the properties of that substance is called a molecule. A molecule can be represented by a formula in terms of the symbols and number of atoms of the elements present in it. This formula is called molecular formula.

Molecular formula may be defined as,

"The symbolic representation of a molecule of any substance describing the actual number of atoms in it, is called its molecular formula." For example, the molecular formula of water is H_2O. Thus, one molecule of water contains two atoms of hydrogen and one atom of oxygen.

The molecular formula of a substance gives the names and actual number of atoms of the various elements present in a molecule of the substance.

Thus, a molecular formula may also be defined as,

"A shorthand notation for the molecule of a substance in terms of the symbols and actual number of atoms of each element present in one molecule of that substance, is called its molecular formula."

The molecular formula of a compound represents one molecule of that substance. Therefore, it represents the molecular mass of the substance. For example, H_2 is the molecular formula of hydrogen. This formula (H_2) represents mass of hydrogen equal to the molar mass of hydrogen: molar mass of hydrogen is 2.016g mol^{-1}.

EMPIRICAL FORMULA

Empirical formula is defined as follows.

The simplest formula of a substance which gives the relative number of atoms of each element present in the molecule of that substance is called as the empirical formula.

For example, a compound contains carbon, hydrogen and oxygen in the ratio 1 : 2 : 1, its empirical formula will be CH_2O. It may be noted that the actual molecule of this compound may contain these elements in any other multiple of the simple ratio expressed by the empirical formula.

The following example will probably make it more clear. A covalent compound has a molecular formula C_6H_6, *i.e.*, there are six carbon atoms and six hydrogen atoms in its molecule. The lowest whole number ratio between the number of carbon and Hydrogen atoms is 1 : 1, (6 : 6 can be simplified to 1 : 1). Therefore, *the empirical formula of a compound having molecular formula of C_6H_6 is CH.*

Diamond has a three-dimensional network (giant) structure in which each carbon atom is bonded to four different carbon atoms. This giant structure may be represented by the formula C_n, where n may have any value from few hundreds to many thousands depending upon the size of the sample. The empirical formula of diamond is C.

Empirical formula of a compound gives the simplest whole-number ratio between the number of atoms of all the elements present in the compound.

Formula mass (also called as the empirical formula mass) of a substance is equal to the sum of the atomic masses of ail the atoms present in the empirical formula. For example, the empirical formula for benzene (C_6H_6) is CH. So,

Formula mass of CH = (12 + 1) u = 13 u (or 13 g/mol)

Relationship Between the Empirical and Molecular Formulae

The molecular formula of a compound is related to its empirical formula as follows.

Molecular formula = n x Empirical formula = (Empirical formula)$_n$

where n may be 1, 2, 3 ....

The value of n can be obtained by the following relationship:

$$n = \frac{\text{Molecular mass}}{\text{Empirical formula mass}}$$

For example, the empirical formula of a compound having molecular mass of 78 is CH.

So,

$$n = \frac{\text{Molecular mass}}{\text{Empirical formula mass}} = \frac{78\text{u}}{(12 + 1)\text{ u}} = \frac{78}{13} = 6$$

Therefore, the molecular formula of the given compound is

$$6 \times (CH) = C_6H_6$$

In case, the value of n is one, then the empirical formula and the molecular formula of the substance are the same.

WRITING CHEMICAL FORMULA OF A COMPOUND

Molecular and empirical formulae of any compound can be determined from the given composition of that compound. This is illustrated below.

The chemical formula of a compound can be determined from the composition of the compound. The composition of a compound is commonly expressed in terms of the percentage of each element present in it.

To determine the formula of an unknown compound follow the following steps.

Step 1. Determine the percentage of each element present in the compound from the mass of each element present in a certain known mass of the compound.

Step 2. If the sum of percentages of all these elements is not 100%, then the difference gives the percentage of oxygen. Calculate the percentage of oxygen in the given compound by using the following relationship.

Percentage of oxygen = 100 - (Sum of the percentages of all other elements present in the compound)

Step 3. Divide the percentage of each element by the atomic mass of the respective element. The ratio so obtained is called atomic ratio.

Step 4. Divide the atomic ratios by the lowest value, and convert these into the nearest whole numbers. These whole numbers give the simplest ratio between the number of atoms of the various elements present in the compound.

Step 5. Write down the empirical formula of the compound.

Step 6. Calculate the empirical formula mass by adding the atomic masses of all the atoms present in the empirical formula.

Step 7. Obtain molecular mass (or molar mass) either from experiment or from the vapour density of the compound by using the relationship,

Molecular mass = 2 × Vapour density

Step 8. Obtain the value of n through the relation,

$$n = \frac{\text{Molecular mass}}{\text{Empirical formula mass}}$$

Step 9. The molecular formula is then obtained by

Molecular formula = n × Empirical formula = (Empirical formula),

The above method of writing the molecular formula of a compound is illustrated through the following numerical problems.

DETERMINATION OF PERCENTAGE COMPOSITION OF A COMPOUND

The mass percentage of each constituent element present in any compound is called its percentage composition.

Experimentally, it can be done by the chemical analysis of the compound. This is done by decomposing a known mass of the given compound, and then estimating each element present therein by a suitable chemical method. Then, the mass percentage of each element can be calculated by using the relationship,

$$\text{Mass percentage of element i} = \frac{\text{Mass of element i}}{\text{Total mass of the compound}} \times 100$$

The percentage composition of a compound may also be obtained from the molecular formula of the compound. This is done as follows.

(a) Write down the molecular formula of the compound.

(b) Calculate the mass of each element present in the molecule of the substance by multiplying the atomic mass of that element by the number of atoms in the molecule.

(c) Calculate the molecular mass of the compound by adding the masses of all the atoms present in its molecule.

(d) Calculate the mass percentage of each element by using the relationship,

Mass percentage of element

$$\text{i} = \frac{\text{Mass of element i in one molecule of the compound}}{\text{Total mass of the compound}} \times 100$$

CHEMICAL EQUATIONS AND THEIR BALANCING

The long-worded statements of any chemical reaction can be summarized by using symbols and formulae of the substances involved in the reaction. Such a representation of a chemical reaction in terms of symbols and formulae is called a chemical equation. A chemical equation thus can be defined as follows:

"A shorthand representation of a chemical reaction in terms of symbols and formulae of the substances involved in the reaction is called a chemical equation."

For example, the reaction of zinc metal with dilute sulphuric acid to produce zinc sulphate and hydrogen may be written as,

Zinc (metal) + dil. Sulphuric acid → Zinc sulphate + Hydrogen

This worded equation may be written in terms of symbols and formulae as,

$$Zn + \text{dil. } H_2SO_4 \rightarrow ZnSO_4 + H_2$$

Thus, $Zn + \text{dil. } H_2SO_4 \rightarrow ZnSO_4 + H_2$, is the chemical equation for the reaction between zinc and dilute sulphuric acid.

Essentials of a Chemical Equation

A chemical equation should satisfy the following conditions:

(i) It should represent a true chemical reaction.

(ii) It should be molecular in nature. The ionic reactions are represented by ionic equations.

(iii) It should be a balanced equation, *i.e.*, the number of atoms of each element on both the sides should be equal.

Writing of a Chemical Equation

Chemical equation for a chemical reaction is written as follows.

Step 1. Identify the reactants and the products of the chemical reaction.

Step 2. Write down the formulae or symbols of the reactants on the left hand side with a sign of plus (+) between them.

The formulae/symbols of the products formed in the reaction are written on the right hand side with a sign of plus (+) between them.

The two sides (reactants and products) are separated either by a sign of equality (=) or that of an arrow (→) pointing towards the products.

For a reversible reaction, a sign ($\rightleftharpoons$) is used in place of arrow.

Such a chemical equation is called the skeleton equation.

Step 3. Count the number of atoms of each element on both the sides. If the number of atoms of each element on both the sides are equal, then the equation is termed as a balanced chemical equation.

If the number of atoms of any one or more of the elements on both the sides are not equal, then these are made equal by adjusting the coefficients before the symbols and formulae of the reactants and products. The process by which the number of atoms of each element on both sides are made equal, is called balancing of chemical equation. The methods for balancing chemical equations are described in details below.

Step 4. In the end, the chemical equation is made molecular, if required.

Information Conveyed by a Chemical Equation

A chemical equation gives the following two types of information.

1. Qualitative information. A chemical equation provides the following qualitative information about the reaction. It tells us the,

   (i) names of the reactants which take part in the reaction.

   (ii) names of the products formed in the reaction.

2. Quantitative information. A chemical equation gives the following quantitative information. It tells us about,

   (i) the number of molecules or atoms of reactants and products taking part in the reaction.

   (ii) the number of moles of each substance involved in the reaction.

   (iii) the mass of each substance involved in the reaction.

   (iv) the mass-mass, mass-volume, volume-volume relationships between the reactants and products.

Limitations of a Chemical Equation

A chemical equation written as suggestive above does not give us the following information. Thus, following are the limitations of a chemical equation.

(i) it does not tell about the reaction conditions, such as temperature, pressure, and the presence of catalyst etc.

(ii) it does not tell about the physical state of the reactants and products.

(iii) it does not tell us whether heat is absorbed or evolved in the chemical reaction.

(iv) it does not tell us anything about the concentration of reactants and products.

(v) it does not tell us anything about the rate of the reaction, *i.e.*, it does not tell us about the time taken by the reaction for its completion.

Making a Chemical Equation More Informative

A chemical equation can be made more informative by adding additional information to the chemical equation, as described below.

(i) *Reaction conditions :* The information regarding temperature, pressure, and catalyst etc., is provided above the arrow (→) or above the sign of equality (=) separating the reactants and products.

A reaction taking place at t°C and p atm pressure, and in the presence of a catalyst can be described as follows.

$$\text{Re actants} \xrightarrow[\text{catalyst}]{\text{t°C, p atm}} \text{Products}$$

For example, nitrogen and hydrogen react together to form ammonia under the conditions; temperature = 450°C, pressure = 200–900 atm, and in the presence of a catalyst (a mixture of iron and molybdenum). The chemical equation for this reaction and giving this information is written as follows.

$$N_2 + 3H_2 \xrightarrow[\text{Fe + Mo}]{\text{450°C, 200–900 atm}} 2NH_3$$

(ii) *Physical states of reactants and products :* Information about the physical state of the reactants and products, can be provided by using the letters (s), (1), (g) and (aq) for solid, liquid, gas and a solution in water, respectively at the end of the formula, or symbol of the substance involved. For example, solid sodium metal reacts with water at room temperature to produce hydrogen gas and a solution hydroxide in water. Then, the complete chemical equation is,

$$2Na(s) + 2H_2O(l) \rightarrow 2NaOH(ag) + H_2(g)$$

The gaseous substance which is evolved (given out) during the reaction may also be indicated by an arrow pointing upwards (↑).

A solid substance which precipitates out from the reaction mixture is indicated by an arrow pointing downwards (↓).

(iii) *Heat absorbed or evolved* : Chemical reactions proceed with the evolution or absorption of heat. The reactions in which heat is absorbed are called *endothermic reactions*. The reactions in which heat is given out are called exothermic reactions. This information is provided by adding a heat term on the product-side (right hand side) of the chemical equation. For example,

(a) When carbon is burnt in air (or oxygen) heat is evolved. Then, the chemical equation is written as,

$$C(s) + O_2(g) \rightarrow CO_2g + \text{Heat (393 kJ)}$$

(b) The reaction between carbon (C) and sulphur (S) to produce carbon disulphide (CS,) proceeds with the absorption of heat, *i.e.*, it is an endothermic reaction. The chemical equation of this reaction is written as,

$$C(s) \mid 2S\ (g) \rightarrow 2CS_2(g) - \text{Heat (92 kJ)}$$

The reactions with + *Heat* term on the product side are called *exothermic reactions*, while those with - *Heat* term on the product side are called endothermic reactions.

The chemical equations in which the quantity of heat absorbed or evolved is included are called *thermochemical equations*.

(iv) *Concentration of the reactants and products* : This information is added to the chemical equation by adding the word dil. (for dilute) or cone. (for concentrated) before the formulae of the reactants and products. For example, in the reaction between zinc and dilute sulphuric acid, the term dil. is added before the formula of sulphuric acid.

$$Zn(s) + \text{dil. } H_2SO_4aq) \rightarrow ZnSO_4aq) + H_2(g)$$

(v) *Rate of reaction* : This information is not commonly added to the chemical equation. Sometimes, however, the term fast or slow may be added over the arrow, if the reaction is fast or slow. For example, the reaction between HCl and NaOH in solution is a fast reaction. So,

$$HCl(aq) + NaOH(ag) \xrightarrow{\text{Fast}} NaCl(ag) + H_2O(l)$$

BALANCING OF CHEMICAL EQUATIONS

The number of atoms of each element on both the sides of the arrow in a chemical equation should be equal. This is because no matter is destroyed or created during a chemical reaction (law of conservation of mass). So, a chemical equation should be balanced so as to satisfy the requirement of the law of conservation of mass.

The method by which the number of atoms of each element on both the sides of the arrow (→) in a chemical reaction are made equal is called balancing of chemical equation.

Balancing of chemical equations is done by any of the following methods

(i) Hit and trial method

(ii) Partial equation method

(iii) Oxidation number method

(iv) Ion-electron method

In this unit, we would describe the first two method. The other two methods are described in Unit 9 of this book.

Hit and Trial Method

This method is also called trial and error method or inspection method.

Balancing of a chemical equation by this method involves the following steps.

Step (i) Write the skeleton equation by writing the symbols and formulae of the reactants and products.

Step (ii) If there is any elementary gas (*e.g.*, 0^, Hp N^ etc.) appearing on either side of the equation, then write it in the atomic state in the skeleton equation.

Step (iii) Start balancing from the formula containing the maximum number of atom: In case, it is not found convenient, then start balancing the atoms which appear the least number of times.

Step (iv) Balance the atoms of any elementary gas in the last+

Step (v) When the balancing is over, convert the equation to the molecular form.

We illustrate this method by balancing the equation for the reaction between methane and oxygen to form carbon dioxide and water.

The reaction is,

$$\text{Reactants} \rightarrow \text{Products}$$

$$\text{Methane} + \text{Oxygen} \rightarrow \text{Carbon dioxide} + \text{Water}$$

Step (i) The skeleton equation is, $CH_4(g) + O(g) \rightarrow CO_2(g) + H_2O(l)$

The elementary gas oxygen is kept in the atomic form.

Step (ii) CH_2 molecule contains the maximum number of atoms. So, start is made from here.

(a) No of C atoms on both the sides are equal.

(b) CH_4 has four H atoms, while there are only 2 H atoms in H_2O on the right side. The H-atoms can be balanced by multiplying H_2O by 2. So, partially balanced chemical equation may be written as,

$$CH_4(g) + O(g) \rightarrow CO_2(g) + 2H_2O(l)$$

Step (iii) Now, the C atoms and H-atoms are balanced. Oxygen atoms can be balanced by multiplying O(g) on the left by 4. The equation then can be written as,

$$CH_4(g) + 4O(g) \rightarrow CO_2(g) + 2H_2O(l)$$

The number of atoms of each element on either side are now equal. So, the equation is balanced.

Step (iv) Converting this equation to molecular form, one can write,

$$CH_4(g) + 2O_2(g) \rightarrow CO_2(g) + 2H_2O(l)$$

This is the balanced chemical equation for the reaction involving the burning of methane.

PARTIAL EQUATION METHOD

Chemical equations involving many reactants and products cannot be balanced easily by the hit and trial method. Such chemical equations are balanced by a method called partial equation method.

In partial equation method, the overall reaction is assumed to take place through two or more simple reactions. Each subtraction can be described by a simple chemical equation. Such simple chemical equations are called partial equations.

To balance a chemical equation by partial equation method follow the following steps.

Step 1. Split the given chemical equation into two or more partial equations (simple equations).

Step 2. Each partial equation is separately balanced by hit and trial method.

Step 3. Such balanced partial equations are multiplied with suitable coefficients so as to exactly cancel out the common substances which do not appear in the overall chemical equation.

Step 4. Add the balanced partial equations obtained in step 3, to obtain the balanced chemical equation.

Application of the partial equation method is illustrated through the following examples.

LIMITING REAGENT

Reactants react in accordance with the stoichiometry indicated by the balanced chemical equation. Quite often, the reacting substances are not present in exactly the same proportions as required by the balanced chemical equation. Some reactants may be present in lesser amounts, while some other may be present in amounts greater than the stoichiometry amounts. The reacting substance which gets used up first in the reaction is called the limiting reagent. This is because the amount of the limiting reagent limits the amount of the products formed. A part of the other reactants which are present in amounts greater than the stoichiometric amounts is left behind as unconsumed reagents. The concept of limiting reagent is illustrated below.

Consider the reaction,

	$2H_2(g)$ +	$O_2(g)$	$\rightarrow$ $H_2O(l)$
Stoichiometric amounts :	2mol	1 mol	2 mol
Here, hydrogen is the limiting reagent:	2 mol	2 mol	2 mol
Here, oxygen is the limiting reagent:	3 mol	1 mol	2 mol

According to the reaction stoichiometry, 2 mol of hydrogen react with 1 mol of oxygen to give 2 mol of water.

Let us consider a reaction mixture consisting of 2 mol of hydrogen and 2 mol of oxygen. According to the stoichiometry, 2 mol of hydrogen require only 1 mol of oxygen. So, in this situation, only 1 mol of oxygen is used up, and 1 mol of oxygen is left behind unreacted. 2 mol of hydrogen are completely consumed producing 2 mol of H_2O. In this case, therefore, hydrogen is the limiting reagent.

Let us consider another situation, Here, 3 mol of hydrogen and 1 mol of oxygen are taken. According to the stoichiometry, 1 mol of oxygen would react with 2 mol of hydrogen to form 2 mol of H_2O. 1 mol of hydrogen is left behind, and the oxygen is completely consumed. In this situation, oxygen is the limiting reagent.

CALCULATIONS USING CHEMICAL EQUATIONS

The relative amounts of reactants and products involved in any reaction are represented by the balanced chemical equation. The balanced chemical equation can be used in many ways exploiting the mass-mass, mass-volume, mole-mass, and volume-volume relationships.

The problems based on these relationships can be solved by following the following steps.

Steps for Solving Problems Based on Chemical Equations

To solve problems based on chemical equations, follow the following steps.

Step 1. Write down the balanced chemical equation.

Step 2. Write down the stoichiometry (no of moles, relative mass/ volume etc.) of each reactant and product below their symbols or formulae.

Step 3. Identify the substances whose quantities are given or to be found out. Write down the actual quantities of the substances given.

Step 4. Calculate the quantity of the desired substance by simple mathematical calculations.

Calculations involving chemical equations are illustrated below.

PROBLEMS BASED ON MASS-VOLUME, AND VOLUME-VOLUME RELATIONSHIP

A few parameters of interest for solving the problems based on the mass-volume and volume-volume relationship are,

Molar volume of a gas at NTP = 22.4 L/mol (= 22400 mL/mol)

1 mole of each gas under NTP conditions occupies a volume of 2.4 L.

These calculations are illustrated through the following examples.

Alternative Method

The above problem may also be solved as follows. The reaction may be written as,

$$\begin{array}{llll} C + & O_2 & \rightarrow & CO_2 \\ 1 \text{ mol} & & & 1 \text{ mol} \\ & & & 22.4 \text{ L} \end{array}$$

Thus, 1 mol of carbon on burning would produce 22.4 L of CO_2 at NTP. We known, that

Molar mass of carbon = 12 g mol^{-1}

Mass of carbon burnt = 1 kg = 1000 g

Therefore,

$$\text{No. of moles of carbon in 1000 g} = \frac{1000 \text{ g}}{12 \text{gmol}^{-1}} = 83.33 \text{ mol}$$

Then;

1 mol of carbon gives = 22.4 L of CO_2

83.33 " " = 83.33 × 22.4 L of CO_2

= 1866.7 L of CO_2

Thus, 1 kg of C on burning gives 1866.7 L of CO_2 at NTP.

CONCENTRATION OF SOLUTIONS AND PROBLEMS BASED ON THEM

Units of Concentration

Most of the chemical reactions are carried out in solutions. So, it is important to know how to express the concentration of a substance in its solution.

Concentration of any solute in a solution may be expressed in many ways. Most common units of concentration are described below.

1. *Mass percent of Percent by mass :* It is defined as the mass of solute in gram per 100 g of the solution. For example, a 5% solution of sodium chloride means that 5 g of NaCl are present in 100 g of the solution. Mathematically, mass percent of a solute in solution can be expressed as,

$$\text{Mass \% of the solute} = \frac{\text{Mass of solute}}{\text{Mass of solution}} \times 100$$

Both, the mass of solute and that of the solution must be expressed in the same mass units, viz., both in grams or both in kilograms etc.

However, when a solute is present in trace quantities (very small quantity) the concentration is expressed in the units of parts per million (abbreviated as ppm). For example, a sample containing 5 g of solute in 10^6 g of a solution is said to have a concentration of 5 ppm. The composition of gaseous mixtures is also expressed in ppm, units based on the volume ratio. For example. 10 mL in 10_6 mL corresponds to 10 ppm.

The atmospheric pollutions generally measured in the units of pollutant in 10^6 units of air.

2. *Volume percent of Percent by volume :* The number of units of volume of the solute per 100 units of volume of the solution is known as volume percent (% v/v). For example, a 5% (v/v) solution of ethyl alcohol contains 5 ML of alcohol in 100 mL of the solution. Mathematically, volume percent is expressed as,

$$\text{Volume percent of solute} = \frac{\text{Volume of solute}}{\text{Volume of the solution}} \times 100$$

3. *Molarity (M) :* Molarity of a solutions defined as the number of moles of solute dissolved per dm^3 (or litre, L) of the solution. Molarity of any solution depends upon temperature. So, molarity of any solution is specified for a given temperature. Mathematically, molarity is defined as,

Molarity of solution

$$= \frac{\text{No.of moles of the solute}}{\text{Volume of the solution in litres}\left(\text{or inde}^3\right)} = \frac{n\,\text{mol}}{V\,L}.$$

where, is the number of moles of the solute,

V is the volume of the solution in liters (or dm^3).

Since, the number of moles of any substance is related to its mass and the molar mass, hence,

$$\text{No. of moles of solute} = \frac{\text{Mass of the solute}}{\text{Molar mass of the solute}}$$

The molarity of a solution then can also be expressed as,

Molarity of the solution

$$= \frac{\text{Mass of the solute}}{\text{Molar mass of the solute} \times \text{Volume of the solution in litres}}$$

So, if W gram of a substance having molar mass M, i dissolved in sufficient solvent, so as to make the total volume of V litre of the solution, then the molarity (M) of the solution is given by

$$\text{Molarity} = \frac{W\ g}{M\ g\ mol^{-1} \times V\ \text{litre}} = \frac{w}{M \times V}\ mol/L$$

In the case of ionic compounds, *e.g.*, Na^+Cl^-, the term molar mass loses its significance, because no definite molecule exists as such. According to Pauling, in such cases the unit of gram formula mass should be used. Thus, in such cases a new concentration unit called formality (F) should be used. Formality is defined as,

"the number of gram formula mass dissolved per litre of the solution is called formality of the solution."

Thus,

$$\text{Formality of the solution} = \frac{\text{No. of formula mass of the solute}}{\text{Volume of the solution in litres}}$$

or Formality of the solution

$$= \frac{\text{mass of the solute}}{\text{Formula mass of the solute} \times \text{Volume of solution in litres}}$$

It may be noted, that the formality of a solution has the same value as molarity of any solution. But, the concept of formality is applied only to the ionic solutes, where no discrete molecule exists.

Both, the molarity and formality have one disadvantage. This is that both the molarity and formality of any solution change with temperature. This is because the solution expands or contracts with the change in temperature.

4. *Normality (N) :* Normality of a solution is defined as the number of equivalents of the solute dissolved per litre (or dm^3) of the solution at any specified temperature. This,

$$\text{Normality of solution} = \frac{\text{No. of equivalents of the solute}}{\text{Volume of the solution in litres}}$$

Since, the number of equivalents of any solute can be obtained from its mass and equivalent mass hence,

Normality

$$= \frac{\text{mass of the solute (W)}}{\text{Equivalent mass of the solute(E)} \times \text{Volume of solution in litres(V)}}$$

or Normality of solution

$$= \frac{W\ g}{E\ g/\text{equiv} \times V\ \text{litre}} = \frac{w}{E \times V}\ \text{equiv}/L$$

Normality of a solution also varies with temperature.

5. *Motility (m) :* Molality of a solution is defined as the number of moles of solute per kg of the solvent. If a solution is prepared by dissolving n moles of a solute in W kg of the solvent, then

Molality (m) of the solution

$$= \frac{\text{No. of moles of solute}}{\text{Mass of the solvent in kg}} = \frac{n_{solute}\ \text{mol}}{W_{solvent}\ \text{kg}}$$

or $$\text{Molality, (m)} \frac{n_{solute}}{W_{solvent}}\ \text{mol kg}^{-1}$$

Since, the number of moles of solute ca be expressed in terms of its and molar mass, hence

$$n_{solute} = \frac{\text{Mass of solute (W)}}{\text{Molar mass of the solute(M)}}$$

So, $$\text{Molarity(m)} = \frac{(w/M)}{W}\ \text{mol kg}^{-1} = \frac{w}{M \times W}\ \text{molkg}^{-1}$$

Depending upon the units of w, M and W, the following two cases become possible.

Case I: If, w is the mass of the solute (in gram units)

m is the molar mass of the solute (in gram per mole units)

W is the mass of the solvent (in kg units)

Then, $\text{Molarity}, (m) = \frac{W\ g}{M\ g\ mol \times W\ Kg} = \frac{w}{M \times W}\ mol/kg$

Case II: w is the mass of the solute (in gram units)

M is the molar mass of the solute (in gram per mole units)

W is the mass of the solvent (in gram units)

Then,

$$\text{Molaity}, (m) = \frac{W\ g}{M\ g\ mol \times W\ g} = \frac{w\ g}{M\ g/mol \times (W/1000) kg}$$

or, $$\text{Molaity}(m) = \frac{1000 w}{M \times W} mol/kg$$

Molality (m) of a solution does not change with temperature.

6. *Molefraction (X) :* The molefraction of any component of a solution is defined as the ratio of the number of moles of that component to the total number of moles of all the components of the solution. Thus, if a solution contained n_A moles of A, and n_B moles of B, then,

Molefraction of A, X_A

$$= \frac{\text{No.of moles of A}}{(\text{No. of moles of A} + \text{No.of moles of B})} = \frac{n_A}{n_A + n_B}$$

Molefraction of B, X_B

$$= \frac{\text{No.of moles of B}}{(\text{No. of moles of A} + \text{No.of moles of B})} = \frac{n_B}{n_A + n_B}$$

Adding the above equations, one gets

$$X_A + X_B = \left(\frac{n_A}{n_A + n_B}\right) + \left(\frac{n_B}{n_A + n_B}\right) = \frac{(n_A + n_B)}{(n_A + n_B)} = 1$$

Thus, the sum of the molefractions of all the components of a solution is equal to one, *i.e.*,

$$X_A + X_B = 1$$

Molefraction does not change with temperature.

Acids like hydrochloric acid, sulphuric acid, nitric acid, acetic acid and bases like ammonium hydroxide are available as concentrated aqueous solutions. The solutions required in the laboratory are prepared by definite w\volume of the concentrated acid by adding a calculated amount of water. If the density of the acid is known, then it is possible to calculate the molarity of the diluted acid solution.

NORMALITY AND MOLARITY EQUATIONS

To dilute a given solution to another of different normality or molarity, we use a mathematical equation called normality (or molarity) equation. The equation is commonly written as,

$$N_1 V_1 = N_2 V_2$$

where, N_1 = Initial normality and N_2 = Final normality

V_1 = Initial volume (of the given solution)

V_2 = Final volume (of the solution to be prepared)

So, if out of these four quantities any three are known then the fourth can be calculated.

When required for the dilution of any solution of a substance, the molarity equation used is written as,

$$M_1 V_1 = M_2 V_2$$

where, M_1 = Initial molarity and M_2 = Final molarity

V_1 = Initial volume (of the given solution)

V_2 = Final volume (of the solution to be prepared)

It may be noted that the molarity equation when used in volumetric titrations is written in the following generalized form,

$$Z_1 M_1 V_1 = Z_2 M_2 V_2$$

where Z_1 is the number of equivalents in one mole of the substance 1

and Z_2 is the number of equivalents in one mole of the substances 2.

CONCEPTUAL CORNER FOR COMPETITIVE EXAMINATIONS

It is an accepted fact that a student learns faster and more efficiently if he ha some sense of participation in the subject he studies. Students studying science realize their sense of participation while conducting experiments in laboratories. Another opportunity to realize this sense of participation comes through solving numerical problems. It is probably this objective that solving numerical problems is an integral part of the curriculum for Science subjects.

The students at this level are generally wary of solving numerical problems and indirect questions. The main objective of this section is not only to provide the student with a large number of numerical problems but also to help then develop problem-solving skills through a step-by-step approach illustrated in the example problems.

Suggestions for Solving Problems Successfully

The following suggestions with help a student to become a successful problem-solver.

* Read the statement of the problem very carefully and determine exactly what is asked for.
* Write down all the given quantities using standard notations and their units.
* Determine the principle and write down the formula required.
* Now proceed in an orderly fashion to perform the necessary mathematical calculations. Report the calculated parameter along with its units.

IMPORTANT FORMULAE

* Vapour density (V.D.)

$$= \frac{\text{Mass of mL of the substance in gaseous state}}{\text{Mass of V mL of hydrogen under similar conditions of T and P}}$$

* Molar mass of the substance = 2 × V.D. of the substance
* $1\text{u} = \frac{\text{Mass of a } {}^{12}_{6}\text{C atom}}{12}$

$$= \frac{1.9924 \times 10^{-23}\ \text{g}}{12} = 1.66 \times 10^{-24}\ \text{g} = 1.66 \times 10^{-27}\ \text{kg}$$

* Relative atomic mass of an element (A_r)

$$= \frac{\text{Average mass of 1 atom of the element}}{\left(\text{Mass of one } {}^{12}_{6}\text{C atom}/12\right)}$$

* Average mass of 1 atom of an element = Relative atomic mass ×

$$\left(\frac{\text{Mass of one } {}^{12}_{6}\text{C atom}}{12}\right)$$

* Atomic mass (A) of an element $= A_r \times 1\ u = A_r\ u$
* Relative molecular mass, M_r

$$= \frac{\text{Average mass of a molecule of the substance}}{\left(\text{Mas of anatom of } {}^{12}_{6}C\right)/12}$$

* Average mass of a molecule of a substance]

$$= \text{Relative molecular mass} \times \frac{\text{Mas of an atom of } {}^{12}_{6}C}{12}$$

* Molecular mass (M) \ $M_r \times 1\ u = M_r\ u$

* The average mass of one mole of any material is called its molar mass. Mathematically,

$$\text{Molar mass (M)} = \frac{\text{Mass of the substance}}{\text{Amount of the substance (in moles)}} = \frac{m}{n}$$

Since, the unit of mass is kg or g, and that of n is s\mol, hence the unit of molar mass is gram per mole (g/mol) or kilogram per mole (kg/mol).

* Molar mass of any substance = Mass of 6.023×10^{23} chemical units of that substance

* Molecular formula = $(\text{Empirical formula})_n$

where $$n = \frac{\text{Molecular formula mass}}{\text{Empirical formula mass}}$$

* Mass percentage of element i in a compound

$$= \frac{\text{Mass of element i}}{\text{Total mass of the compound}} \times 100$$

*Mass percentage of element in a compound]

$$= \frac{\text{Mass of element i in one molecule of the compound}}{\text{Molecular mass of the compound}} \times 100$$

* Mass percentage of the solute in the solution

$$= \frac{\text{Mass of solute}}{\text{Mass of solution}} \times 100$$

* Volume percent of solute in the solution

$$= \frac{\text{Volume of solute}}{\text{Volume of solution}} \times 100$$

* No of moles of solute $= \dfrac{\text{Mass of the solute}}{\text{Molaer mass of the solute}}$

* Molarity of solution

$$= \frac{\text{No. of moles of the solute}}{\text{Volume of the solution in litres (or in dm}^3)} = \frac{n}{V}\ \text{mol L}^{-1}$$

* Molarity of the solution

$$= \frac{\text{Mass of the solute / molar mass of the solute}}{\text{Volume of the solution in litres}}$$

* Molarity $= \dfrac{W\ g / M\ g\ mol^{-1}}{V\ \text{litre}} = \dfrac{W}{M \times V}$ mol/ L

* Normality of solution $= \dfrac{\text{No.of equivalents of the solute}}{\text{Volume of the solution in litres}}$

* Normality

$$= \frac{\text{Mass of the solut (W) / Equivalent mass of the solute (E)}}{\text{Volume of the solution in litres (V)}}$$

* Normality of solution $= \dfrac{W g / E g\ \text{equiv}^{-1}}{V\ \text{litre}} = \dfrac{W}{E \times V}$ equiv / L

* Molality (m) of the solution

$$= \frac{\text{No. of moles of solute}}{\text{Mass of the solvent in kg}} = \frac{n_{solute}\ \text{mol}}{W_{solvent}\ \text{kg}} = \frac{n_{solute}}{W_{solvent}}\text{mol kg}^{-1}$$

$$n_{solute} = \frac{\text{Mass of solute (w)}}{\text{Molarmss of solute (M)}}$$

$$\text{Molality (m)} = \frac{(w/M)}{W}\text{mol kg}^{-1} = \frac{w}{M \times W}\ \text{mol kg}^{-1}$$

Molarity equation. The generalized form of the molarity equation is

$$z_1 M_1 V_1 = z_2 M_2 V_2$$

where z_1 is the number of equivalents in one mole of the substance 1

and, z_2 is the number of equivalents in one mole of the substance 2

SOLVED EXAMPLES

Example 1:

Carbon and oxygen are known to form two compounds. The carbon content in one of these is 42.9% while in the other it is 27.3%. Show that this data is in agreement with the law of multiple proportions.

Solution:

For first compound

Mass % of C = 42.9

$\therefore$ Mass % of O = 57.1

Thus, 42.9 g of C reacts with 57.1 g of oxygen

$$1\text{g of C reacts with } \frac{57.1}{42.9} \text{ g oxygen}$$

$= 1.33$ g of oxygen

For second compound

Mass % of C = 27.3

$\therefore$ Mass % of O = 72.7

Thus, 27.3 g of C reacts with 72.7 g of oxygen

$$1\text{g of C reacts with } \frac{72.7}{27.3} \text{ g of oxygen}$$

$= 2.66$ g of oxygen

The ratio of oxygen masses which combine with 1 g of C is

1.33 : 2.66 or 1 : 2

Since, 1 : 2 is a simple ratio, hence the law of multiple proportions is supported by the data.

Example 2:

Hydrogen sulphide contains 94.11% sulphur dioxide contains 50% oxygen. Water contains 11.11% hydrogen. Show that the results are in agreement with the law of reciprocal proportions.

Solution:

(a) Calculation of masses of sulphur and oxygen that combine with certain fixed mass, say 1 g of hydrogen,

(i) In H_2S, 100 – 94.11 = 5.89 g hydrogen combines with 94.11 g sulphur. So,

1 g hydrogen combines with $\frac{94.11}{5.89}$ g = 15.98 g of sulphur

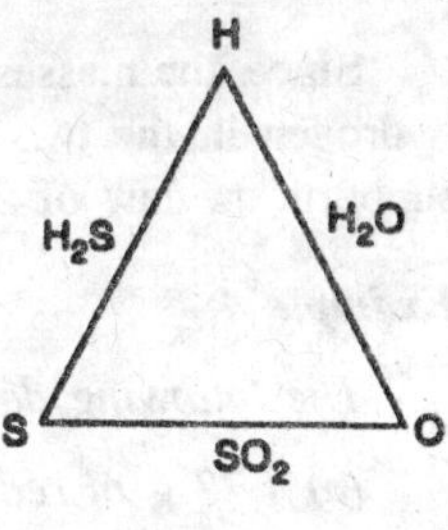

Fig. 1

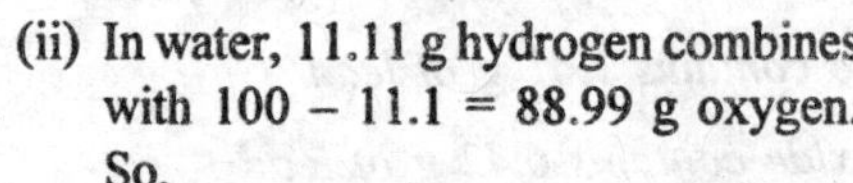

(ii) In water, 11.11 g hydrogen combines with 100 – 11.1 = 88.99 g oxygen. So,

$$1 \text{ g hydrogen combines with } \frac{88.89}{11.11} \text{ g} = 8 \text{ g oxygen}$$

(b) Calculation of ratio of the masses of sulphur (in H_2S) and oxygen (in H_2O)

$$\frac{\text{Mass of sulphur}}{\text{Mass of oxygen}} = \frac{15.98}{8} = 2 \qquad \text{...(i)}$$

(c) Calculation of ratio of the masses of sulphur and oxygen when they combine to form sulphur dioxide.

$$\frac{\text{Mass of sulphur}}{\text{Mass of oxygen}} = \frac{50}{50} = 1 \qquad \text{...(ii)}$$

The ratio (i) is double of (ii), *i.e.*, 2 : 1, which is a simple ratio. This illustrates the law of reciprocal proportions.

Example 3:

Hydrogen peroxide and water contain 5.93% and 11.2% of hydrogen respectively. Show that the data illustrates the law of multiple proportions.

Solution:

The given data can be tabulated as follows.

Compound	***Hydrogen (%)***	***Oxygen(%) (calculated)***	***Mass of oxygen per unit mass of hydrogen***	***Simple mass ratio of oxygen***
Hydrogen peroxide	5.93	100–5.93=94.07	94.07/5.93 = 15.86	15.86/7.93=2
Water	11.2	100–11.2=88.8	88.8/11.2=7.93	7.93/7.93=1

Since, the masses of oxygen which combine with one unit mass of hydrogen in the two compounds are in the ratio 2:1, hence the data supports the law of multiple proportions.

Example 4:

The following data were obtained for three oxides of lead:

(a) 1.77 g of red oxide contains 1.61 g of lead

(b) 6.90 g of yellow oxide contains 6.42 g of lead

(c) 7.795 g of brown oxide contains 1.035 g of lead

Show that the data supports the law of multiple proportions.

Solution:

The given data is tabulated as follows.

Oxide	*Given mass(g)*		*Calculated mass of oxygen*	*Mass of oxygen per gram of lead*	*Ratio of different masses of O_2 per gram of lead*	*Simple ratio*
	Oxide	*Lead*	*(g)*			
1. Red	1.77	1.61	0.16	0.16/1.61 = 0.1	0.1/0.075 = 1.33	4
2. Yellow	6.90	6.42	0.48	0.48/6.42 = 0.075	0.075/0.075 = 1.0	3
3. Brown	1.195	1.035	0.16	0.16/1.035 = 0.15	0.15/0.075 = 2.0	6

Thus, the masses of oxygen which combine with the same mass (1g) of lead are in a simple ratio. So, the given data supports the law of multiple proportions.

Example 5:

Calculate the number of atoms in each of the following :

(i) 52 moles of He

(ii) 52 u of He

(iii) 52 g of He

Solution:

(i) 1 mole of He contains 6.023×10^{23} He atom

52 " " $\frac{6.023 \times 10^{23} \times 52}{1}$ He atom

$= 3.132 \times 10^{25}$ He atom

(ii) 1 molecule of He ≡ 4 u

$\therefore$ $52 \text{ u} = 1 \times \frac{52\text{ u}}{4\text{ u}}$ He atom = 13 He atom

(iii) We have,

1 mol of He ≡ 4 g ≡ 6.023×10^{23} He atom

Thus, 4 g of He contains 6.023×10^{23} He atom

52 g " " $6.023 \times 10^{23} \times \frac{52}{4}$ He atom

$= 7.83 \times 10^{24}$

He atom.

Example 6:

Calculate the mass of the following :

(a) 1 atom of carbon

(b) 1 atom of silver

(c) 1 molecule of benzene (C_6H_6)

(d) 1 molecule of water (H_2O)

Solution:

(a) Molecular mass of carbon = 12.011 u

So, Molar mass of carbon = 12.011 g mol^{-1}

No. of carbon atoms in one mole = 6.023×10^{23} mol^{-1}

Therefore,

$$\text{Mass of 1 atom of carbon } \frac{12.011 \text{ g mol}^{-1}}{6.023 \times 10^{23} \text{ mol}^{-1}} = 1.99 \times 10^{-23} \text{g}$$

(b) Molecular mass of silver = 107.87 u

So, Molar mass of silver = 107.87 g mol^{-1}

Then,

$$\text{Mass of 1 atom of silver} = \frac{\text{Molar mass}}{\text{Avogadro's number}}$$

$$= \frac{107.87 \text{ g mol}^{-1}}{6.023 \times 10^{23} \text{ mol}^{-1}} = 17.91 \times 10^{-23} \text{ g}$$

(c) Molecular mass of benzene (C_6H_6) = (6 × 12.01 u) + (6 × 1 u)

= 72.06 u + 6 U

= 78.06 u

So, Molar mass of benzene = 78.06 g mol^{-1}

Then, Mass of 1 molecule of benzene = $\frac{\text{Molar mass of benzene}}{\text{Avogadro's number}}$

$$= \frac{78.06 \text{ g mol}^{-1}}{6.023 \times 10^{23} \text{ mol}^{-1}} = 12.94 \times 10^{-23} \text{ g}$$

(d) Molecular mass of water = (2 × 1 u) + (1 × 16 u)

= 2u + 16 u = 18 u

So, Molar mass of water = 18 g mol^{-1}

Then,

Mass of 1 molecule or water = $\frac{\text{Molar mass of water}}{\text{Avogadro's no.}}$

$$= \frac{18 \text{ g mol}^{-1}}{6.023 \times 10^{23} \text{ mol}^{-1}}$$

= 2.99 × 10^{-23} g

Example 7:

One million silver atoms weigh 1.79 × 10^{-16} g. Calculate the atomic mass of silver.

Solution:

No. of silver atoms = 1 million = 1 × 10^6

Mass of one million Ag atoms = 1.79 × 10^{-16} g

Mass of 6.023 × 10^{23} atoms of silver

$$= \frac{1.79 \times 10^{-16}}{1 \times 10^{6}} \times 6.023 \times 10^{23} = 107.8 \text{ g}$$

Atomic mass of silver is equal to the mass of 6.023 × 10^{23} atoms of silver. So, the atomic mass of silver is 107.8. u.

Example 8:

(a) 0.2 mol of NH_3

(b) 3.0 mol of CO_2

(c) 5.14 mol of H_5IO_6

Solution:

(a)	Molar mass of NH_3	= (1 × 14.0 + 3 × 1.0) g/mol	= (14 + 3) g/mol	= 17 g/mol
Then,	Mass of 0.2 mol of NH_3	= 0.2 mol × 17 g/mol	= (0.2 × 17) g	= 3.4 g
(b)	Molar mass of CO_2	= (1 × 12.0 + 2 × 16.0)g/mol	= (12 + 32) g/mol	= 44 g/mol
Then,	Mass of 3.0 mol of CO_2	= 3.0 mol × 44 g/mol	= 132 g	
(c)	Molar mass of H_5IO_6	= (5 × 1)+(1 × 127)+(6 × 16)g/mol		
		= (5 + 127 + 96) g/mol	= 228 g/mol	
Then,	Mass of 5.14mol of H_5IO_6	= 5.14 mol × 228 g/mol	= 1171.9 g	=1.172 kg

Example 9:

A sample of gaseous substance weighing 0.5 g occupies on volume of 1/12 litre under NTP conditions. Calculate molar mass of substance.

Solution:

1 mole of any gaseous substance at NTP occupies 22.4 L. So,

1.12 L of gaseous substance = 0.5 g

1 L " " $= \frac{0.5}{1.12 \text{ L}}$ g

22.4 L " " $= \frac{0.5 \text{ g} \times 22.4 \text{ L}}{1.12 \text{ L}} = 10$ g

The molar mass of the substance therefore is 10 g/mol.

Example 10:

Calculate number of atoms of each type in 5.3 g of Na_2CO_3.

Solution:

Molar mass of Na2CO_3 = (2×12.0)+(1×12.0)+(3 × 16.0)g mol^{-1}

= (46.0+12.0+48.0g mol^{-1}=106.0 g mol^{-1}

So, Number of moles of Na_2CO_3 in 5.3 g of Na_2CO_3

$$= \frac{5.3 \text{ g}}{106.0 \text{ mol}^{-1}} = 0.05 \text{ mol}$$

From stoichiometry

Na_2CO_3	≡	2Na	+	C	+	3 O
1 mol		2 mol		1 mol		3 mol
0.05 mol		2 × 0.05 mol		1 × 0.05 mol		3 × 0.05 mol
		= 0.1 mol		= 0.05 mol		= 0.15 mol

Since, one mole of any substance contains 6.02×10^{23} chemical units, hence

No. of Na atoms in 5.3 g of $NaCO_3$ = $0.1 mol \times 6.023 \times 10^{23} mol^{-1}$

$= 6.023 \times 10^{23}$

No. C atoms in 5.3 g of Na2CO3 = $0.05 mol \times 6.023 \times 10^{23} mol^{-1}$

$= 3.011 \times 10^{22}$

No. of O atoms in 5.3 g of Na2CO3 = $0.15 mol \times 6.023 \times 10^{23} mol^{-1}$

$= 9.034 \times 10^{22}$

Example 11:

How many years would it take to spend Avogadro number of rupees at the rate of Rs. 10 lakh per second?

Solution:

We know,

Avogadro's number = 6.023×10^{23}

So, Total money to be spent = 6.023×10^{23} rupees

Time taken to spend Rs. 10 lakh (= 10^6) = 1 s

Therefore,

Time taken to spend 6.023×10^{23} rupees

$$= \frac{1s}{10^6} \times 6.023 \times 10^{23} = 6.023 \times 10^{17} \text{ s}$$

The time in seconds my be converted into years as follows.

$$6.023 \times 10^{17} \text{ s} = \frac{6.023 \times 10^{17}}{60 \times 60 \times 24 \times 365} \text{ yr} = 1.91 \times 10^{10} \text{ yr}$$

So, the time taken to spend Avogadro's number of rupees is 1.91 × 10^{10} year.

Example 12:

What would be the mass of 5.0 mole of NH_3? Calculate the number of NH_3 molecules and nitrogen and hydrogen atoms in it. (Avogadro's number is 6.022×10^{23}).

Solution:

Molecular mass of NH_3 ≡ (Atomic mass of N) + 3 (Atomic mass of hydrogen).

or Molecular mass of NH_3 = 14.0 u + (3 × 1.008 u) = 17.024 u

So, 5.0 mol of NH_3 = 5 mol × 17.0 g mol–1 = 85.0 g

So, the mass of 5 mol of ammonia is 85.0 g.

From the mole concept, we know that one mole of any ;gas contains Avogadro's number of molecules. Therefore,

NH_3	≡	N	+	3 H
1 mol		1 mol		3 mol
5 mol		5 mol		15 mol
$(5 \times 6.023 \times 10^{23})$		$(5 \times 6.023 \times 10^{23})$		$(15 \times 6.023 \times 10^{23})$
30.1×10^{23} molecule		30.1×10^{23} atom		90.3×10^{23} atom

So, 5 mol of ammonia contain,

(a) 30.1×10^{23} molecule of ammonia

(b) 30.1×10^{23} atom of nitrogen

(c) 90.3×10^{23} atoms of hydrogen

Example 13:

How may moles of each kind of atoms are there in one mole of each of the following ?

(i) HCl (ii) H_2SO_4 (iii) $BaCl_2$

(iv) $AlCl_3$ (v) PCl_5

Solution:

(i) HCl is the molecular formula for hydrochloric acid gas. One molecule of this compound is made up of 1 hydrogen atom and 1 atom of chlorine, that is

HCl	≡	H		Cl
1 molecule	≡	1 atom		1 atom
1 mol	≡	1 mol		1 mol

Thus,

1 mol of HCl ≡ 1 mol of H atom + 1 mol of Cl atom

Therefore, one mole of HCl contains 1 mole of H-atoms and 1 mole of Cl-atoms

Similarly, we can write,

(ii) 1mol of H_2SO_4 ≡ 2 mol of H-atoms + 1 mol of S-atom + 4 mol of O-atom

(iii) 1mol of $BaCl_2$ ≡ 1 mol of Ba-atom + 2 mol of Cl atom

(iv) 1mol of $AlCl_3$ ≡ 1 mol of Al-atom + 3 mol of Cl atom

(v) 1mol of PCl_5 ≡ 1 mol of P-atom + 5 mol of Cl-atom

Example 14:

Chlorophyll, the green colouring matter of plants responsible for photosynthesis, contains 2.68% of magnesium by weight. Calculate the number of Mg atoms in 2.00 g of chlorophyll.

Solution:

100 g of chlorophyll contains 2.68 of Mg

$\therefore$ 2 g " " $\frac{2.68}{100} \times 2 = 5.36 \times 10^{-2}$ g of Mg

Form the mole concept,

1 mol of Mg ≡ 24 g of Mg ≡ 6.023×10^{23} atom of Mg

So, 24 g of Mg contains 6.023×10^{23} atom of Mg

1 g " " $\frac{6.023 \times 10^{23}}{24}$

5.36×10^{-2} g " $\frac{6.023 \times 10^{23} \times 5.36 \times 10^{-2}}{24}$

$= 1.34 \times 10^{21}$ Mg atom

Example 15:

The cost of table salt (NaCl) and table sugar ($C_{12}H_{22}O_{11}$) is Rs. 2 per kg and Rs. 6 per kg respectively. Calculate their cost per mole.

Solution:

(i) Molar mass of NaCl $= (23 + 35.5)$ g/mol $= 58.5$ g/mol

Cost of table salt per 1000 g = Rs. 2

So, Cost of 58.5 g table salt $= \frac{\text{Rs.}2 \times 58.2}{1000} = \text{Rs. } 0.117 = 12 \text{ paise}$

So, one mole of table salt would cost 12 paise.

(ii) Molar mass of table sugar ($C_{12}H_{22}O_{11}$)

$= (12 \times 12) + (22 \times 1) + (11 \times 16)$ g/mol

$= (144 + 22 + 176)$ g/mol

$= 342 \text{ g mol}^{-1}$

Cost of table sugar per 1000 g = Rs. 6

So, Cost of 342 g of table sugar $= \frac{\text{Rs.}6 \times 342}{1000}$

$= \text{Rs. } 2.05$

So, one mole of sugar would cost Rs. 2.05.

Example 16:

Calculate the mass of carbon monoxide having the same number of oxygen atoms as are present in 88 g of carbon dioxide.

Solution:

Oxygen balance in carbon monoxide and carbon dioxide can be expressed by the relationship,

CO_2	≡	2 CO
1 mol		2 mol
(12 + 32)g		2(12 + 16)g
44 g		56 g
88 g		$\frac{56 \text{ g} \times 88 \text{g}}{44 \text{ g}} = 112 \text{ g}$

So, 112 g of carbon monoxide contain the same number of oxygen atoms as present in 88 g of carbon dioxide.

Example 17:

Determine the percentage composition of potassium nitrate, (KNO_3).

(Atomic masses (in atomic mass unit) are; K = 39, N = 14, O= 16).

Solution:

Formula mass of KNO_3 = (39 u + 14 u + 3 × 16 u) = 101 u

Then,

$$\text{Mass percentage of potassium} = \frac{39 \text{ u}}{101 \text{ u}} \times 100 = 38.6$$

$$\text{Mass percentage of nitrogen} = \frac{14 \text{ u}}{101 \text{ u}} \times 100 = 13.9$$

$$\text{Mass percentage of oxygen} = \frac{3 \times 16 \text{ u}}{101 \text{ u}} \times 100 = 47.5$$

The percentage composition of KNO_3 is K= 38.6%, N = 13.9% and O = 47.5%.

Example 18:

Sugar ($C_{12}H_{22}O_{11}$) costs Rs. 15 per kg. How much one mole of sugar would cost ?

Solution:

Molar mass of sugar = (12 × 12.0 + 22 × 1.0 + 11 × 16.0) g/mol

= (144 + 22 + 176) g/mol = 342 g/mol

1 kg (= 1000 g) of sugar costs = Rs. 15

$$1 \text{ g} \quad " \quad = \frac{\text{Rs } 15}{1000}$$

$$342 \text{ g} \quad " \quad = \frac{342 \times \text{Rs. } 15}{1000}$$

$$= \frac{\text{Rs.} 15 \times 342}{1000} = \text{Rs. } 4.72$$

Example 19:

Calculate the number of water molecules present in a spherical drop of water having a radius of 1 mm. Take density of water as 1 g/cm³.

Solution:

Radius of the spherical water drop = 1 mm = 0.1 cm

Volume of one drop of water

$$= \frac{4}{3}\pi r^3 = \frac{4}{3} \times \frac{22}{7} \times (0.1)^3 \text{ cm}^3 = 4.19 \times 10^{-3} \text{ cm}^3$$

So, Mass of water in one drop = Volume × Density

$= 4.19 \times 10^{-3}$ cm^3 × 1 g cm^{-3}
$= 4.19 \times 10^{-3}$ g

Molar mass of water (H_2O)

= (2 × 1.0 + 1 × 16.0) g/mol

= 18 g/mol

So, 18 g of water contain $= 6.023 \times 10^{23}$ molecules

1 g " " $= \frac{6.023 \times 10^{23}}{18}$

4.19×10^{-3} g " " $= \frac{4.019 \times 10^{-3} \times 6.023 \times 10^{23}}{18}$

molecules $= 1.4 \times 10^{20}$ molecules

Example 20:

A compound contains 75% carbon and 25% hydrogen. Determine its empirical formula. The molecular mass of this compound is 16 u. Determine its molecular formula also. The atomic masses are : C = 12 u, H = 1 u.

Solution:

The above results are written as follows :

Element	*Percentage composition*	*Atomic mass*	*Atomic ratio*	*Simplest atomic ratio*
C	75	12	75/12 = 6.25	6.25/6.25 = 1
H	25	1	25/1 = 25	25/6.25 = 4

So,

Empirical formula of the compound = C_1H_4 or CH_4

Then,

Empirical formula mass = (1 × 12 u) + (4 × 1 u)

= 12 u + 4 u = 16u

Molecular mass (given) = 16 u

So, $n = \frac{\text{Molecular mass}}{\text{Empirical formula mass}} = \frac{16u}{16u} = 1$

Therefore, Molecular formula= 1 × Empirical formula

$= 1 \times CH_4 = CH_4$

Example 21:

A substance on analysis gave the following percentage composition:

Na = 43.4%, C = 11.3% and O = 45.3%. Determine its empirical and molecular formulae.

Given: the relative molecular mass of the compound is 106.

Solution:

Given results are tabulated a follows :

Element	*Percentage composition*	*Atomic mass*	*Atomic ratio*	*Simplest atomic ratio*
Na	43.4	23	43./23 = 1.89	1.89/0.94 = 2
C	11.3	12	11.3/12 = 0.94	0.94/94 = 1
O	45.3	16	45.3/16 = 2.83	2.83/0.94 = 3

Thus, the simplest ratio Na, C and O in the given compound is 2 : 1 : 3. Hence, the empirical formula of the given compound is Na_2CO_3.

Then,

Empirical formula mass of the compound

= (2 × 23 u) + (1 × 12 u) + (3 × 16 u)

= 46 u + 12 u + 48 u = 106 u

Then, $n = \frac{\text{Molecular mass}}{\text{Empirical formula mass}} = \frac{106u}{106u} = 1$

Therefore, molecular formula of the given compound is given by,

Molecular formula = 1 × Empirical formula

$= 1 \times Na_2CO_3 = Na_2CO_3$

Example 22:

A compound was found to have 78.2% boron and 21.8% of H. Its molar mass was determined to be 27.6 g mol^{-1}. What is the molecular formula of the compound ? Atomic masses are boron (B) - 11 u, and hydrogen (H) = 1 u.

Solution:

The given data can be tabulated as follows :

Element	*Percentage composition*	*Atomic mass*	*Atomic ratio*	*Simplest atomic ratio*
Boron (B)	78.2	11	78.2/11 = 7.1	7.1/7.1 = 1
Hydrogen (H)	21.8	1	21.8/1 = 21.8	21.8/7.1 = 3

Thus, the empirical formula of the compound is BH_3.

Empirical formula mass = (1 × atomic mass of B) + (3 × atomic mass of hydrogen)

$$= (1 \times 11) + (3 \times 1)\ u = 14\ u$$

Molecular mass = 27.6 u

So, $$n = \frac{\text{Molecular mass}}{\text{Empirical formula mass}} = \frac{27.6u}{14u} = 1.97$$

Since, n can have only integral values, hence n = 2. So,

Molecular formula of the compound = 2 × Empirical formula = 2 × BH_3 = B_2H_6

The molecular formula of the given compound is B_2H_6.

Example 23:

An organic compound was found to contain the following constituents : C = 33.8%, H = 4.7%, N = 13.2%, Cl = 33.4% and the rest is oxygen. Determine its empirical formula.

Solution:

Mass % of oxygen = 100 – (33.8 + 4.7 + 13.2 + 33.4) = 14.9

The entire analytical data is tabulated in the following table.

Element	*Mass %*	*Atomic mass*	*Atomic ratio*	*Simplest atomic ratio*
C	33.8	12	33.8/12 = 2.8	2.8/093 = 3
H	4.7	1	4.7/1 = 4.7	4.7/0.93 = 5
N	13.2	14	13.3/14 = .094	0.94/0.93 = 1
Cl	33.4	35.5	33.4/35.5 = 0.94	0.94/0.93 = 1
O	14.9	16	14.9/16 = 0.93	0.93/0.93 = 1

The empirical formula is C_3H_5NClO

Example 24:

A compound is found to contain 11.2% nitrogen (N), 3.2% hydrogen (H), 41.2% chromium (Cr) and 44.4% oxygen (O). Determine the stoichiometric (empirical) formula. Atomic masses (amu) are : N = 14, H = 1, Cr = 52 and O = 16.

Solution:

The above result are tabulated as follows.

Element	*Percentage composition*	*Atomic mass*	*Atomic ratio*	*Simplest atomic ratio*
N	11.2	14	11.2/14 = 0.8	0.8/0.8 = 1 or 2
H	3.2	1	3.2/1 = 3.2	3.2/0.8 = 4 or 8
Cr	41.2	52	41.2/52 = 0.8	0.8/0/8 = 1 or 2
O	44.4	16	44.4/16 = 2.8	2.8/0.8 = 3.5 or 7

Since, the simplest atomic ratios cannot have fractional values, hence the simplest atomic ratio of N, H, Cr and O are 2, 8, 2, 7 respectively. Therefore,

The stoichiometric (empirical) formula of the compound = $N_2H_8Cr_2O_7$

(The actual formula of this compound is $(NH_4)_2\ Cr_2O_7$. This compound is named as ammonium dichromate. Ammonium dichromate contains NH_4^+ and $Cr_2\ O_7^{2-}$ ions).

Example 25:

Write the empirical formulae of the compounds having the following molecular formulae.

(i) C_6H_6 (ii) C_6H_{12} (iii) H_2O_2 (iv) H_2O (v) Na_2CO_3 (vi) B_2H_6 (vii) N_2O_4

Solution:

(i) CH (ii) CH_2 (iii) HO (iv) H_2O (v) Na_2CO_3 (vi) BH_3 (vii) NO_2

Example 26:

(a) Butyric acid contains C, H and O. A 4.24 mg same of butyric acid is completely burnt. It gives 8.45 mg of CO_2 and 3.46 mg of H_2O. What is the mass % of each element in butyric acid ?

(b) If the elemental composition of butyric acid is found to be 54.2% C, 9.2% H and 36.6%O, determine its empirical formula.

(c) The molecular mass of butyric acid was determined to be 88 u. What is the molecular formula ?

Solution:

Form the given data

(a) Mass % of C in butyric acid

$$= \frac{12\,g}{44\,g} \times \frac{\text{Mass of } CO_2}{\text{Mass of butyric acid}} \times 100$$

$$= \frac{12}{44} \times \frac{8.45}{4.24} \times 100 = 54.3$$

Mass % of H in butyric acid

$$= \frac{12}{44} \times \frac{\text{Mass of } H_2O}{\text{Mass of butyric acid}} \times 100$$

$$= \frac{2}{18} \times \frac{3.46}{4.24} \times 100 = 9.07$$

Then, Mass % of O = 100 – (54.3 + 9.07) = 36.6

(b)

Element	*Mass %*	*Atomic mass*	*Atomic ratio*	*Simplest atomic ratio*
C	54.2	12	54.2/12 = 4.5	4.5/2.3 = 2
H	9.2	1	9.2/1 = 9.2	9.2/2.3 = 4
O	36.6	16	36.6/16 = 2.3	2.3/2.3 = 1

Therefore, the empirical formula of butyric acid is C_2H_4O.

(c) Molecular mass of butyric acid = 88 u

Empirical formula mass

$$= (2 \times 12 \text{ u}) + (4 \times 1 \text{ u}) + (16 \text{ U})$$

$$= 24 \text{ u} + 4 \text{ u} + 16 \text{ u} = 44 \text{ u}$$

$$\text{Therefore, } n = \frac{\text{Molar mass}}{\text{Empirical formula mass}} = \frac{88\text{u}}{44\text{u}} = 2$$

So, Molecular formula of butyric acid = $2 \times C_2H_4O = C_4H_8O_2$

Example 27:

An organometallic compound on analysis gave the following results. C = 64.4%, H = 5.5% and Fe = 29.9%. Determine its empirical formula.

Solution:

The given data is tabulated as follows. Since, the sun of mass % is 99.8, hence there is no oxygen in the given compound.

Element	***Mass %***	***Atomic mass***	***Atomic ratio***	***Simplest ratio***	***Simplest whole number ratio***
C	64.4	12	64.4/12 = 5.37	5.37/053 = 10.1	10
H	5.5	1	5.5/1 = 5.5	5.5/0.53 = 10.4	10
Fe	29.9	56	29.9/56 = 0.53	0.53/0.53 = 1	1

Thus, the empirical formula of the compound is $C_{10}H_{10}Fe$.

Example 28:

Acetylene burns in oxygen to form carbon dioxide and water. Write the skeleton equation for the reaction, and balance the chemical equation by hit and trial method.

Solution:

Step (i) The skeleton equation is,

$$C_2H_4(g) + O(g) \rightarrow H_2O(l) + CO_2(g)$$

The elementary gas oxygen is written in the atomic form.

Step (ii) Starting from the formula having the maximum number of atoms, *i.e.*, $C_2H_2(g)$. Inspect on of the skeleton equation shows that H-atoms on the sides are equal. C-atoms can be made equal by multiplying $CO_2(g)$ on the right side by 2. So, the partially balanced equation is,

$$C_2H_2(g) + O(g) \rightarrow H_2O(l) + 2CO_2(g)$$

Step (iii) Now, there are 5 O-atoms on the right, and only one on the left. Therefore, O-atoms can be balanced by multiplying O (g) on the left by 5. The resulting equation is,

$$C_2H_2(g) + 5O(g) \rightarrow H_2O(l) + 2CO_2(g)$$

Step (iv) Now, the equation is balanced. Convert this atomic equation to the molecular equation by multiplying throughout by 2. Thus,

$$2C_2H_2(g) + 5O_2(g) \rightarrow 2H_2O(l) + 4CO_2(g)$$

This is the balanced equation.

Example 29:

A compound of titanium and potassium was found to contain k = 23.1%; Ti = 14.2 % and Cl = 62.7 %. Establish its empirical formula.

Solution:

Since the sum of mass % of K, Ti and Cl is 100, there is no oxygen in the compound. The given data are processed as follows :

Element	*Mass %*	*Atomic mass*	*Atomic ratio*	*Simplest ratio*	*Simplest whole number ratio*
K	23.1	39.1	23.1/39.1=0.592	0.592/0.297=1.99	2
Ti	14.2	47.9	14.2/47.9=0.297	0.297/0.297=1	1
Cl	62.7	35.5	62.7/35.5=1.77	1.77/0.297=5.96	6

Therefore, the empirical formula of this compound is K_2TiCl_6.

Example 30:

A compound contains 92.3% carbon and 7.7% hydrogen. Its molecular mass is 18.0 u. Establish its empirical and molecular formula.

Solution:

The sum of mass % of C And H is 100. So there is no oxygen in the compound. Now to establish the empirical formula the data are processed as follows.

Element	*Mass %*	*Atomic mass*	*Atomic ratio*	*Simplest ratio*	*Simplest whole number ratio*
C	92.3	12	92.3/12 = 7.69	7.69/7.69 = 1	1
H	7.7	1	7.7/1 = 7.7	7.7/7.69 = 1	1

Therefore, the empirical formula of this compound of this compound is CH.

To determine molecular formula:

Molecular mass (given) = 78 u

Empirical formula mass = (12 u + 1 u) = 13 u

So,
$$n = \frac{\text{Molecular formula mass}}{\text{Empirical formula mass}}$$

$$= \frac{\text{Molecular mass}}{\text{Empirical formula mass}} = \frac{78\text{ u}}{13\text{ u}} = 6$$

Therefore, Molecular formula= n × Empirical formula

$$= 6 \times (CH) = C_6H_6$$

Example 31:

40 g of copper chloride was analyses. The following results were obtained .

Mass of copper = 1.890 g; Mass of chlorine = 2.10 g. Determine the empirical formula of copper chloride.

Solution:

From the given data,

$$\text{Mass \% of Cu} = \frac{1.890}{40} \times 100 = 47.25$$

$$\text{Mass \% of Cl} = \frac{2.110}{4.0} \times 100 = 52.75$$

Element	*Mass %*	*Atomic mass*	*Atomic ratio*	*Simplest ratio*	*Simplest whole number ratio*
Cu	47.25	63.5	47.25/63.5=0.74	0.74/0.74 = 1	1
Cl	52.75	35.5	52.75/35.5=1.48	1.48/0.74 = 2	2

So, the empirical formula of copper chloride is $CuCl_2$.

Example 32:

Rewrite the following equation in the blanched form showing in it that Al $(OH)_3$ is an insoluble product.

$$Al_2(SO_4)_3 + NaOH \rightarrow Al(OH)_3 + Na_2SO_4$$

Solution:

(i) The inspection of the equation,

$$Al_2(SO_4)_3 + NaOH \rightarrow Al(OH)_3 + Na_2SO_4$$

Shows that there is no elementary gas involved in this reaction. So, balancing is started from the molecule $Al_2(SO_4)_3$ (containing the maximum number of atoms).

(ii) There are 2Al atoms on the left and only 1 on the right. So, Al atoms can be balanced by multiplying $Al(OH)_3$ (on the right) by 2, *i.e.*,

$$Al_2(SO_4)_3 + NaOH \rightarrow 2Al(OH)_3 + Na_2SO_4$$

(iii) Now, there are six OH groups on the right, and only 1 on the left. The – OH groups can be balanced by multiplying NaOH by 6. So, the chemical equation can be written as,

$$Al_2(SO_4)_3 + 6NaOH \rightarrow 2Al(OH)_3 + Na_2SO_4$$

(iv) Now, Al atoms and OH groups are balance. To balanced Na atoms and SO_4 groups multiply Na_2SO_4 on the right by 3. The chemical equation now become s

$$Al_2(SO_4)_3 + 6NaOH \rightarrow 2Al(OH)_3 + 3Na_2SO_4$$

This is the balanced equation. $Al(OH)_3$ is an insoluble product. This information is included in the equation by adding either (s) or an arrow (↓) pointing downward immediately after $Al(OH)_3$. So, the balanced chemical equation, showing that $Al(OH)_3$ is an insoluble product is,

$$Al_2(SO_4)_3 + 6NaOH \rightarrow 2Al(OH)_3\ (s) + 3Na_2SO_4$$

or $$Al_2(SO_4)_3 + 6NaOH \rightarrow 2Al(OH)_3\ (\downarrow) + 3Na_2SO_4$$

Example 33:

Balance the chemical equation,

$$KMnO_4 + HCl \rightarrow KCl + MnCl_2 + H_2O + Cl_2$$

by hit and trial method.

Solution:

(i) Skeleton equation : $KMnO_4 + HCl \rightarrow KCl + MnCl_2 + H_2O + Cl_2$

(ii) Writing elementary gas in atomic form : $KMnO_4 + HCl$

$$\rightarrow KCl + MnCl_2 + H_2O + Cl$$

(iii) Starting from $KMnO_4$: K and Mn are balanced.

(iv) Balancing O atoms : $KMnO_4 + HCl$

$$\rightarrow KCl + MnCl_2 + 4H_2O + Cl$$

(v) Balancing H atoms : $KMnO_4 + 8HCl$

$$\rightarrow KCl + MnCl_2 + 4H_2O + Cl$$

(vi) Balancing Cl atoms : $KMnO_4 + 8HCl$

$$\rightarrow KCl + MnCl_2 + 4H_2O + 5Cl$$

This is balanced atomic equation.

(vii) Making it molecular : $2KMnO_4 + 16HCl$

$$\rightarrow 2KCl + 2MnCl_2 + 8H_2O + 5Cl_2$$

So, the balanced molecular equation is,

$$2KMnO_4(aq) + 16HCl(aq)$$

$$\rightarrow 2KCl(aq) + 2MnCl_2(aq) + 8H_2O(l) + 5Cl_2(g)$$

Example 34:

Balance the following equation :

(i) $H_3PO_2 \rightarrow H_3PO_4 + PH_3$

(ii) $Ca + H_2O \rightarrow Ca(OH)_2 + H_2$

(iii) $Fe(SO_4)_3 + NH_3 + H_2O \rightarrow Fe(OH)_3 + (NH_4)_2SO_4$

Solution:

(i) The given equation : $H_3PO_2 \rightarrow H_3PO_4 + PH_3$

Balancing of P on both the sides requires a coefficient of 2 before H_3PO_2

Thus, $2H_3PO_2 \rightarrow H_3PO_4 + PH_3$

(ii) $Ca + H_2O \rightarrow Ca(OH)_2 + H_2$

First, balance Ca atoms (already balanced)

Then, balance H atoms–this requires multiplication of H_2O by 2, hence

$$Ca + 2H_2O \rightarrow Ca(OH)_2 + H_2$$

The O atoms get balanced in the process. Therefore, balanced equation is

$$Ca + 2H_2O \rightarrow Ca(OH)_2 + H_2$$

(iii) $Fe_2(SO_4)_3 + NH_3 + H_2O \rightarrow Fe(OH)_3 + (NH_4)_2SO_4$

(a) Balance Fe atoms – it requires multiplication of $Fe(OH)_3$ by 2. Thus, the partly balanced equation is

$$Fe_2(SO_4)_3 + NH_3 + H_2O \rightarrow 2Fe(OH)_3 + (NH_4)_2SO_4$$

(b) Balance SO_4^{2-} ion – it require 3 before $(NH_4)_2SO_4$. The resulting equation is

$$Fe_2(SO_4)_3 + NH_3 + H_2O \rightarrow 2Fe(OH)_3 + 3(NH_4)_2SO_4$$

(c) To balance N atoms multiply NH_3 by 6. The resulting equation is,

$$Fe_2(SO_4)_3 + 6NH_3 + H_2O \rightarrow 2Fe(OH)_3 + 3(NH_4)_2SO_4$$

(d) H an d O atoms can be balanced by multiplying H_2O by 6. The completely balanced equation is,

$$Fe_2(SO_4)_3 + 6NH_3 + 6H_2O \rightarrow 2Fe(OH)_3 + 3(NH_4)_2SO_4$$

Example 35:

Balance the equation, $NaOH + Cl_2 \rightarrow NaCl + NaClO_3 + H_2O$ by partial equation method.

Solution:

(i) Skeleton equation, $NaOH + Cl_2 \rightarrow NaCl + NaClO_3 + H_2O$, can be spit into two partial equations.

Partial eq. 1 : $NaOH + Cl_2 \rightarrow NaCl + NaClO + H_2O$

Partial eq. 2 : $NaClO \rightarrow NaClO_3 + NaCl$

(ii) Balancing the two partial equations by hit and trial method gives,

Balanced partial eq. 1 : $2NaOH + Cl_2 \rightarrow NaCl + NaClO + H_2O$

Balanced Partial eq. 2: $3\ NaClO \rightarrow NaClO_3 + 2\ NaClO$

(iii) NaClO does not appear in the equation for the overall reaction. So, to exactly cancel it, the balanced partial equation 1 is multiplied by 3. Then the matched, balanced partial equations and the balanced over al equation are,

$$2\ NaOH + Cl_2 \rightarrow NaCl + NaClO + H_2O\ ] \times 3$$

$$3\ NaClO \rightarrow NaClO_3 + 2\ NaCl$$

$$6\ NaOH + 3Cl_2 \rightarrow 5NaCl + NaClO_3 + 3H_2O$$

(Balanced chemical equation)

Example 36:

$KMnO_4$ in an acidic solution, oxidizes ferrous sulphate to ferric sulphate. Write the skeleton equation for this reaction and balance it by partial equation method .

Solution:

(i) The skeleton equation is,

$$KMnO_4 + H_2SO_4 + FeSO_4 \rightarrow K_2SO_4 + MnSO_4 + Fe_2(SO_4)_3 + H_2O$$

(ii) This reaction is assumed to proceed through the following reactions .

Partial equation 1 : $KMnO_4 + H_2SO_4 \rightarrow K_2SO_4 + MnSO_4 + H_2O + O$

Partial equation 2 : $FeSO_4 + H_2SO_4 + O \rightarrow Fe_2(SO_4)_3 + H_2O$

Following the principle of hit and trial method, the two equations given above are balanced and are suitably multiplied to cancel out the common element, as shown below :

Balanced Partial eq 1 : $2KMnO_4 + 3H_2SO_4 \rightarrow K_2SO_4 + 2MnSO_4 + 3H_2O + 5O$

Balanced Partial eq 2 : $2FeSO_4 + H_2SO_4 + O \rightarrow Fe_2(SO_4)_3 + H_2O] \times 5$

Balanced over-equation $2KMnO_4 + 8H_2SO_4 + 10FeSO_4 \rightarrow K_2SO_4 + 2MnSO_4 + 5Fe_2(SO_4)_3 + 8H_2O$

Example 37:

Write the skeleton equation for the reaction between acidified potassium permanganate and oxalic acid. Balance the skeleton equation by the partial equation method.

Solution:

The reactants and products in this reaction are,

Reactants : Potassium permanganate ($KMnO_4$), Sulphuric acid (H_2SO_4), Oxalic acid ($H_2C_2O_4$).

Products : Potassium sulphate (K_2SO_4), Manganese sulphate ($MnSO_4$), Water (H_2O) and Carbon dioxide (CO_2).

Then, the skeleton equation is,

$$KMnO_4 + H_2SO_4 + H_2C_2O_4 \rightarrow K_2SO_4 + MnSO_4 + H_2O + CO_2$$

(i) The overall reaction is split into the two reactions represented by the following partial equations.

Partial eq 1 : $KMnO_4 + H_2SO_4 \rightarrow K_2SO_4 + MnSO_4 + H_2O + O$

Partial eq 2 : $H_2C_2O_4 + O \rightarrow 2CO_2 + H_2O$

(ii) These partial equations are balanced by hit and trial method as illustrated above. The balanced partial equations are,

Balanced Partial eq 1 : $2KMnO_4 + 3H_2SO_4 \rightarrow K_2SO_4 + 2MnSO_4 + 3H_2O + 5O$

Balanced Partial eq 2 : $H_2C_2O_4 + O \rightarrow 2CO_2 + H_2O$

(iii) Balancing the above balanced partial equations with respect to each other, *i.e.*, by balancing the number of O atoms evolved and consumed, one gets,

$$2\,KMnO_4 + 3H_2SO_4 \rightarrow K_2SO_4 + 2MnSO_4 + 3H_2O + 5O \quad ...(a)$$

$$H_2C_2O_4 + O \rightarrow 2CO_2 + H_2O\ ] \times 5 \quad ...(b)$$

(iv) The balanced chemical equation for the overall reaction is obtained by adding the above equations (Eqs. (a) and (b).

$$2\,KMnO_4 + 3H_2SO_4 \rightarrow K_2SO_4 + 2MnSO_4 + 3H_2O + 5O$$

$$H_2C_2O_4 + O \rightarrow 2CO_2 + H_2O\ ] \times 5$$

$$2KMnO_4 + 3H_2SO_4 + 5H_2C_2O_4 \rightarrow K_2SO_4 + 2MnSO_4 + 8H_2O + 10CO_2$$

Example 38:

3.00 g of H_2 react with 29.00 g of O_2 to yield H_2O.

(i) Which is the limiting reactant ?

(ii) Calculate the maximum amount of H_2O that can be formed.

(iii) Calculate the amount of one of the reactants which remains unreacted.

Solution:

The reaction is,	H_2	+	O_2	→	H_2O
Mass of reactants :	3.00 g		29.00 g		
Moles of reactants :	$\frac{3.00\text{g}}{2.016\text{g/mol}} = 1.49$ mol		$\frac{29.00\text{g}}{32.0\text{g/mol}} = 0.906$ mol		

The balanced equation is,

$2H_2$	+	O_2	→	$2H_2O$
2 mol		1 mol		2 mol
1 mol		1/2 mol		1 mol
1.49 mol		$\frac{1.49}{2}$ mol = 0.745 mol		1.49 mol

Thus, 1.49 mol of hydrogen reacts with 0.745 mol of oxygen Since, in this reaction, 0.906 mol of oxygen is taken, hence whole of the hydrogen is used up. Therefore,

(i) Hydrogen is the limiting reagent.

(ii) The maximum amount of water (product) that can be formed is 1.49 mol, (equal to the amount of hydrogen gas)

(iii) Since, oxygen is taken in slight excess, hence oxygen is left is the system as unreacted.

Amount of oxygen left unreacted = 0.906 mol – 0.745 mol = 0.161 mol.

Example 39:

2 atoms of hydrogen (H) combine with one atom of oxygen (O) to give one molecule of water (H_2O). How many grams of hydrogen are needed completely with 6.4 g of oxygen gas ?

Solution:

The chemical reaction is,

$$2H + O \rightarrow H_2O$$

2 atom	1 atom	1 molecule
2 mol	1 mol	1 mol
2 × 1.0 g	1 × 16.0 g	[molar mass of H atom = 1.0g/mol; molar mass of O atom = 16.0 g/mol]
2 g	16 g	
?	6.4 g	

So,

$$\text{Mass of hydrogen required} = \frac{2\text{ g}}{16\text{ g}} \times 6.4\text{ g} = 0.8\text{ g}$$

Example 40:

When zinc sulphide (ZnS) is strongly heated in excess of air, zinc oxide (ZnO) is formed, and gaseous SO_2 is evolved. Calculate the mass of ZnO and SO_2 that can be obtained from 4.866 g of ZnS.

Solution:

The reaction is written as,

2 ZnS	+	$3O_2$	→	2ZnO	+	$2SO_2$
2 mol		3 mol		2 mol		2 mol
2[65+32]g		3[2×16]g		2[65+16]g		2[32+32]g
2 × 97 g		3 × 32 g		2 × 81 g		2 × 64 g
194 g		96 g		162 g		128 g

Thus,

194 g of ZnS gives = 162 g of ZnO

$$1\text{ g} \quad '' \quad '' = \frac{162}{194}$$

$$4.866\text{ g of ZnS gives} = \frac{4.866 \times 162}{194} = 4.06\text{ g ZnO}$$

Similarly,

194 g of ZnS gives = 128 g of SO_2

$$1 \quad '' \quad '' = \frac{128}{194}$$

$$4.866\text{ g of ZnS gives} = \frac{4.866 \times 128}{194} = 3.21\text{ g } SO_2$$

Therefore, 4.866 g of ZnS gives 4.06 g of ZnO and 3.21 g of SO_2.

Example 41:

What mass of N_2 will be required to produce 34.0 g of NH_3 by the reaction.

$$N_2 + 3H_2 \rightarrow 2NH_3 \ ?$$

Solution:

The reaction is,

N_2	+	$3H_2$	$\rightarrow$	$2NH_3$
1 mol		3 mol		2 mol
(2 × 14.0) g				2 × (1 × 14.0 + 3 × 1.0) g
28.0 g				34.0 g

Thus, to produce 34.0 g of ammonia (NH_3), 28 g of nitrogen is required.

Example 42:

Methane burns ;in oxygen to form carbon dioxide and water. Write the balance chemical equation for this reaction. From the above reaction, calculate the mass of oxygen required for burning 1.6 g of methane. (C = 12, 0 = 16, H = 1)

Solution:

The balanced equation for the reaction is

CH_4	+	$2O_2$	$\rightarrow$	CO_2	+	$2H_2O$
(1 × 12.0 + 4 × 1.0)g		2 × (2 × 16.0) g				
(12 + 4) g		64.0 g				
16.0 g						

16.0 g of methane requires = 64.0 g oxygen

1 g " " $= \frac{64.0}{16}$ g Oxygen

1.6 g " " $= \frac{64.0 \times 1.6}{16}$ g oxygen = 6.4 g oxygen

Example 43:

What mass of copper oxide will be obtained by heating 12.35 g of copper carbonate ? Atomic mass of copper = 63.5 u.

Solution:

The reaction may be represent by the chemical equation,

	$CuCO_3$	$\xrightarrow{heat}$	CuO	$+ CO_2(g)$
Stoichiometric mass :	(63.5 + 12 + 48) g		(63.5 + 16) g	
	123.5 g		79.5 g	
Actual mass :	12.35 g		?	

Now, we can write,

123.5 g of $CuCO_3$ gives = 79.5 g CuO

1 " " $= \frac{79.5}{123.5}$

12.35 " " $= \frac{12.35 \times 79.5}{123.5}$ g of $CuCO_3$

$= 7.95$g of $CuCO_3$

Therefore, the mass of copper oxide formed = 7.95 g.

Example 44:

Copper (Cu) wire reacts with silver nitrate ($AgNO_3$) solution to give silver metal (Ag) and a solution of copper nitrate ($Cu(NO_3)_2$). How many gram of metallic silver is produced, if 2.37 g of copper has reacted ?

Solution:

The balanced chemical equation for the reaction is,

$$Cu + 2AgNO_3(aq) \rightarrow 2Ag + Cu(NO_3)_2$$

1 mol = 63.5 g 2 mol=2 ×108g

= 216 g

So,

63.5 g of copper produces = 216 g of silver

1 g " " $= \frac{216 \text{ g}}{63.5}$ "

2.37 g " " $= \frac{216 \times 2.37}{63.5}$ g of silver = 8.06 g

Example 45:

Hydrazine, (N_2H_4) and hydrogen peroxide (H_2O_2), are used together as a rocket fuel. The products of the reaction are nitrogen and water.

How many grams of H_2O_2 are needed per kg (1000 g) of hydrazine carried by the rocket ?

Solution:

The reaction can be represented by the equation,

$$N_2H_4 \quad + \quad 2H_2O_2 \quad \rightarrow N_2 + 4H_2O$$

Stoichiometric mass :(2 × 14 + 4 × 1.0) 2 × (2 × 1.0 + 2 × 16.0)

(28 4) = 32 g 2 × (34) = 68 g

Thus,

32 g of N2H4 require = 68 g of H_2O_2

1 g " " $= \frac{68}{32}$ g of H_2O_2

1000g " " $= \frac{1000 \times 68}{32}$ g of H_2O_2 = 2125 g of H_2O_2 = 2.125 kg of H_2O_2

Example 46:

Find the percentage purity of sodium chloride, 6.5 g of which when dissolved in water and treated with excess of silver nitrate solution gave 14.35 g of silver chloride.

Solution:

Mass of the given sample (impure) of sodium chloride = 6.5 g

Mass of silver chloride formed = 14.35 g

Percentage purity of sodium chloride = ?

The chemical equation for the reaction is,

$$NaCl + AgNO_3 \rightarrow AgCl(s) + NaNO_3$$

Stoichiometric mass :(3+35.5)g (108+35.5)g

58.5 g 143.5 g

Thus,

143.5 g of silver chloride is obtained from = 58.5 g of pure NaCl

1 g " " " $= \frac{58.5}{143.5}$

14.35 g " " " $= \frac{14.35 \times 58.5}{143.5} g$

$= 5.85$ g

∴ Mass of pure NaCl in 6.5 g of impure sample = 5.85 g

Percentage purity of sodium chloride $= \frac{58.5}{6.5} \times 100 = 90$

Example 47:

KBr (potassium bromide) contains 32.9 % by weight potassium. If 6.40 g of bromine reacts with 3.60 g potassium, calculate the number of moles of potassium which combine with bromine to form KBr.

Solution:

Here, 6.40 g of bromine and 3.60 g of potassium are made to react,

Mass % of potassium in reaction mixture

$$= \frac{3.60 \text{ g}}{(6.40 + 3.60)\text{g}} \times 100 = \frac{3.60}{10} \times 100 = 36.0$$

Since, KBr contains 32.9 % (by mass) potassium, hence in the reaction mixture potassium is in slight excess. Therefore, whole of the bromine (6.40 g) would react, and a part of potassium would remain unreacted.

The mass of potassium reacting with 6.40 g of bromine can be found as follows .

Let x g of potassium react with 6.40 g of bromine. Then

Mass % of potassium in KBr $= \frac{x}{6.40 + x} \times 100$

KBr contains 32.9 % of potassium by mass. So,

$$\frac{100\,x}{6.40 + x} = 32.9$$

$$x = 3.14 \text{ g}$$

Thus, only 3.14 g of potassium reacts with 6.40 g of bromine.

Molar mass of potassium = 39g/mol

So,

Number of moles of potassium which combine with 6.4 g of bromine

$$= \frac{3.14\ g}{39\ g/mol} = 0.08\,mol$$

Example 48:

2.3 g of metallic sodium reacts with excess of water. Calculate the mass of sodium hydroxide formed. What is the volume of hydrogen at NTP.

Solution:

$2Na(s)$	+	$2H_2O$	→	$NaOH(aq)$	+	H_2
2 mol				2 mol		1 mol
2 × 23 g				2 (23 + 16 + 1)g		2 g
46 g				80 g		22.4 L (at STP)

Thus,

46 g of sodium gives = 80 g of NaOH

1 " " $= \frac{80}{46}$

2.3 " " $= \frac{2.3 \times 80}{46} = 4.0$ g NaOH

and

46 g of sodium gives = 22.4 L of H_2 gas

1 " " $= \frac{22.4}{46}$

2.3 " " $= \frac{22.4 \times 2.3}{46} = 1.12$ L

Example 49:

Calculate the volume of oxygen at NTP obtained by decomposing 12.26 g of $KClO_3$.

Solution:

$KClO_3$ decomposes as follows,

$$2KClO_3(s) \xrightarrow{heat} 2KCl\,(s) + 3O_2(g)$$

2 mol		3 mol
2(39 + 35.5 + 48) g		3 × 22.4 L (at NTP)
(2 × 122.5) = 245 g		67.2 L (at NTP)

Thus,

245 g of $KClO_3$ give = 67.2 L of oxygen

1 " " $= \frac{67.2}{24.5}$ L of oxygen

12.26 g " " $= \frac{12.26}{24.5} \times 67.2$ L of oxygen

= 3.36 L of oxygen

Therefore, volume of oxygen liberated = 3.36 L

Example 50:

What volume of CO_2 under NTP conditions will be obtained by completely burning 1 kg of carbon in an excess of oxygen ?

Solution:

The reaction between carbon and oxygen can be written as,

	C	+	O_2 (excess)	→	CO_2 1 mol
Stoichiometric mass :	12 g				22.4 L at NTP
Given mass :	1000g				?

So,

12 g of carbon gives = 22.4 L of CO_2

1g " " $= \frac{22.4}{12}$

1000 g " " $= \frac{1000 \times 22.4}{12}$ L of CO_2

= 1866.7 L of CO_2

Thus, 1 kg of carbon on burning would give 1866.7 L of CO_2 under NTP conditions.

Example 51:

What volume of air containing 21 % of oxygen (by volume) is required to completely burn 10 g of sulphur which has a purity level of 98% ?

Solution:

Mass of 98% pure sulphur = 10 g

Volume of air = ?

Percentage of oxygen in air = 21% (by volume)

From the given data,

Mass of pure sulphur in 10 g sample = $\frac{98}{100} \times 10 = 9.8$ g

The reaction may be written as,

$$S + O_2 \rightarrow SO_2$$

1 mol = 32 g   1 mol = 22.4 L at NTP

Thus,

For 32 g of S, one requires = 22.4 L of O_2

1 g " " $= \frac{22.4}{32}$

9.8 " " $= \frac{9.8 \times 22.4}{32} = 6.89$ L of O_2

Thus, 9.8 g of pure sulphur (or 10 g of the given sample of S) requires 6.86 L of oxygen at NTP for complete combustion . We are given,

Air → Oxygen
100 L   21 L

Therefore, For 21 L of oxygen one requires = 100 L of air

1 L " " $= \frac{100}{21}$ L of air

6.86 L " " $= \frac{6.86 \times 100}{21}$ L of air at NTP

= 32.7 L of air at NTP

Example 52:

A small piece of commercial zinc weighing 10 g is made to react with excess of dil. sulphuric acid. The total volume of hydrogen gas liberated was found to be 3.1 litre at NTP. Determine the percentage purity of the zinc sample.

Solution:

Mass of zinc sample = 10 g

Volume of H_2 liberated at NTP = 3.1 L

Percentage purity of zinc = ?

The reaction may be written as,

$$\underset{1\,\text{mol} = 65\text{ g}}{Zn} + \text{dil. } H_2SO_4 \rightarrow ZnSO_4 + \underset{1\text{ mol}=22.4\text{L(at NTP)}}{H_2}$$

Then,

22.4 L of H_2 gas is liberated from = 65 g of zinc

1 L " " $= \frac{65}{22.4}$

3.1 L " " $= \frac{3.1 \times 65}{22.4}$ 8.99 g of zinc

Thus, 10g of commercial zinc sample contains 8.99 of pure zinc. Therefore,

Percentage purity of zinc $= \frac{8.99}{10} \times 100 = 89.9$

Example 53:

(a) A sample of NaOH weighing 0.38 g s dissolved in water and the solution is made to 50.0 mL in a volumetric flask. What is the molarity of the resulting solution?

(b) How many moles of NaOH are contained in 27 mL of 0.15 M NaOH solution ?

Solution:

(a) Molar mass of NaOH = 40 g mol^{-1}

Mass of NaOH dissolved = 0.38 g

No. of moles of NaOH $= \frac{0.38}{40}$ mol

Volume of the solution $= \frac{50\,\text{mL}}{1000\text{ mL/L}} = \frac{50}{1000}$ L

Molarity of the solution

$$= \frac{\text{No.of moles}}{\text{Volume in litres}} = \frac{(0.38/40)}{(50/1000)} = 0.19\,\text{mol}\,L^{-1}$$

(b) From the equation used in (a).

No. of moles of NaOH = Molarity of solution × Volume in litres

$= 0.15\text{ mol } L^{-1} \times 27/1000\text{ L} = 0.00040\text{ mol} = 40 \times 10^{-4}\text{ mol}$

Example 54:

Calculate the molality of a 1 M solution of sodium nitrate. The density of the solution is 1.25 g cm^{-3}.

Solution:

Molar mass of sodium nitrate ($NaNO_3$) = (23+14+48)g/mol

= 85 g/mol

Mass of 1 dm^3 (or 1 litre) of the solution = Volume × Density

= 1000 cm^3×1.25g/cm^3

= 1250 g

Therefore,

Mass of the water containing 85 g of $NaNO_3$ = (1250 – 85)g

= 1165g = 1.165kg

So, Molality (m) of the solution $= \frac{1\,\text{mol}}{1.165\ \text{kg}} = 0.86\ \text{mol kg}^{-1}$

Example 55:

What is the molality of ammonia in a solution containing 0.85 g of NH_3 in 100 cm^3 of a liquid of density 0.85 g cm^{-3} ?

Solution:

Mass of ammonia, $w_2 = 0.85$ g

Molar mass of ammonia, $M_2 = 17$ g mol^{-1}

Mass of liquid (solvent), $W_1 = 100$ cm^3 × 0.85g cm^{-3} = 85 g

So, Molality of ammonia

$$= \frac{n_{NH_3}}{\text{Mass of solvent in kg}} = \frac{w_2 / M_2}{85/1000} = \text{mol kg}^{-1}$$

or, Molality of ammonia

$$= \frac{0.85/17}{(85/1000)}\ \text{mol kg}^{-1} = 0.59\ \text{mol kg}^{-1}$$

Example 56:

Calculate the molefraction of water in a mixture of 12 g water, 108 g acetic acid, and 92 g ethyl alcohol.

Solution:

Following the procedure of the previous problem, one can write

$$n(H_2O) = \frac{\text{Mass of water}}{\text{Molar mass of water}} = \frac{12\,g}{18\,g\,mol^{-1}} = 0.67\,mol$$

$$n(C_2H_5OH) = \frac{\text{Mass of ethanol}}{\text{Molar mass of ethanol}} = \frac{92\,g}{46\,g\,mol^{-1}} = 2.00\,mol$$

$$n(CH_3COOH) = \frac{\text{Mass of acetic acid}}{\text{Molar mass of acetic acid}} = \frac{108\,g}{60\,g\,mol^{-1}} = 1.80\,mol$$

So,

Total number of moles in the solution, n_{total} = (0.67 + 2.00 + 1.80) mol = 4.47 mol

Therefore, $$X_{water} = \frac{n(H_2O)}{n_{total}} = \frac{0.67\text{ mol}}{4.47\text{ mol}} = 0.15$$

$$X_{ehanol} = \frac{n(C_2H_5HO)}{n_{total}} = \frac{2.00\text{ mol}}{4.47\text{ mol}} = 0.45$$

$$X_{acetic\ acid} = \frac{n(CH_3COOH)}{n_{total}} = \frac{1.80\text{ mol}}{4.47\text{ mol}} = 0.40$$

Example 57:

What volume of 95% sulphuric acid (density = 1.85 g/cm³) and what mass of water must be taken to prepare 100 cm³ of 15% solution of sulphuric acid (density = 1.10 g/cm³)?

Solution:

Volume of the solution = 100 cm³

Density of the solution = 1.10 g cm^{-3}

Therefore, Mass of 100cm³ of solution = 100 × 1.10 g = 110 g

The given solution is 15% .This means that 100 g of solution contains 15 g of H_2SO_4.

Mass of H_2SO_4 in 110 g (= 100 cm³) of solution

$$= \frac{15\,g}{100\,g} \times 110\text{ g} = 16.5\text{ g}$$

So, Mass of water in 110 g (= 100 cm³) of solution = (110 – 16.5) g = 93.5 g

So, to obtain 100 cm^3 of 15% solution acid, we require

Mass of water = 93.5 g

Mass of H_2SO_4 (100% pure) = 16.5 g

Since, the given sulphur acid is 95% pure, hence

$$\text{Mass of } H_2SO_4 \text{ (95\%) required} = \frac{100 \times 16.5\text{ g}}{95} = 17.37\text{ g}$$

Density of 95% H_2SO_4 = 1.85 g cm^{-3}

Therefore,

Volume of 95% sulphuric acid required

$$= \frac{\text{Mass}}{\text{Density}} = \frac{17.37\text{ g}}{1.85\text{ g cm}^{-3}} = 9.39\text{ cm}^3$$

Example 58:

A 1.84 g mixture of calcium carbonate and magnesium carbonate upon heating gave 0.96 g residue. Calculate the percentage composition of the mixture. (Atomic masses are: Ca = 40.0 u, Mg = 24.0 u, C = 12.0 u, O = 16.0 U)

Solution:

Calcium, carbonate ($CaCO_3$) and magnesium carbonate ($MgCO_3$) decomposes on heating to give CaO and MgO respectively.

$$CaCO_3 \xrightarrow{\text{heat}} CaO + CO_2$$

(40 + 12 + 48) g (40 + 16) g

= 100 g = 56 g

$$MgCo_3 \xrightarrow{\text{heat}} MgO + CO_2$$

(24 + 12 + 48) g (24 + 16) g

= 84 g = 40 g

Let,

Mass of $CaCO_3$ in the mixture = xg

Then, Mass of $MgCO_3$ in the mixture = (1.84 – x)g

100 g of $CaCO_3$ gives = 56 g of CaO

1 g " " $= \frac{56}{100}$ g of CaO

x g " " $= \frac{x \times 56}{100}$ g of CaO

and 84 g of $MgCO_3$ gives = 40 g of MgO

1 g " " $= \frac{40}{84}$ g of MgO

(1.84 – x) g " " $= (1.84 - x) \times \frac{40}{84}$ g of MgO

So, Total mass of the residue

$$= \frac{x \times 56}{100} \text{ g} + \frac{(1.84 - x) \times 40}{84} \text{ g}$$

$$0.96 = \frac{x \times 56}{100} + \frac{(1.84 - x)40}{84}$$

or x = 1 g

So, Mass of $CaCO_3$ in the mixture = 1 g

Mass of $MgCO_3$ in the mixture = (1.84 g – 1 g)

= 0.84 g

This gives,

Mass % of $CaCO_3$ in the mixture $= \frac{1 \text{ g}}{1.84 \text{ g}} \times 100 = 54.4\ \%$

Mass % of $MgCO_3$ in the mixture $= \frac{0.84 \text{ g}}{1.84 \text{ g}} \times 100 = 45.6\ \%$

Example 59:

An hourly energy requirement of an astronaut can be satisfied by the energy released when 34 grams of sucrose ($C_{12}H_{22}O_{11}$) are burnt in his body. How many grams of oxygen would be need to be carried in space capsule to meet his requirement for one day ?

Solution:

The reaction taking place inside the body is,

$$C_{12}H_{22}O_{11} + 12O_2 \rightarrow 12\ CO_2 + 11H_2O$$

(12×12+22×1+11×16) 12 (2 × 16)

(144 + 22 + 176)g 12 × 32 g

342 g 384 g

So, Mass of oxygen needed for 1 hour

$$= \frac{384 \text{ g}}{342 \text{ g}} \times 34 \text{ g} = 38.17 \text{ g}$$

Then, Mass of oxygen needed for one day = 38.17 g × 24 = 916.2 g

Example 60:

Commercially available concentrated hydrochloric acid contains 38% HCl by mass.

(a) What is the molarity of this solution? The density is 1.19 g.mL.

(b) What volume of conc. HCl is required to make 1.00 L of 0.10 M HCl?

Solution:

(a) 38% solution, means 38 g of HCl in 100 g of solution.

Then, Mass of the solution = 100 g

∴ Volume of the solution

$$= \frac{100 \text{ g}}{\text{Density}} = \frac{100 \text{ g}}{1.19 \text{ g/mL}} = 84.0 \text{ mL} = \frac{84.0}{1000} \text{ L}$$

Molar mass of HCl = 36.5 g mol^{-1}

$$\text{No. of moles of HCl dissolved} = \frac{38}{365} \text{ mol} = 1.04 \text{ mol}$$

∴ Molarity of HCl solution

$$= \frac{1.04 \text{ mol}}{(84.1000) \text{ L}} = 12.38 \text{ mol L}^{-1}$$

(b) Molarity of conc. HCl sample = 12.38 mol/L

Molarity of HCl solution to be prepared = 0.10 mol/L

Volume of HCl solution to be prepared = 1.00 L = 1000 mL

Then, using molarity equation, $M_1V_1 = M_2 V_2$

$$V_{HCl} = \frac{0.10 \times 1000}{12.38} \text{ mL} = 8.08 \text{ mL}$$

Thus, to obtain 1.0 L of 0.10 M HCl, one should dissolve 8.08 mL of conc. HCl to make up the volume to 1.0 L.

Example 61:

Calculate the molality of 1 liters solution of 93% H_2SO_4 (weight/volume). The density of the solution is 1.84 g mL^{-1}.

Solution:

Volume of 93% of H_2SO_4 = 1 litre = 1000 mL

Molar mass of H_2SO_4 = (2 + 32 + 64) g mol^{-1} = 98 g mol^{-1}

Density of the solution = 1.84 g mL^{-1}

Therefore, Mass of 1 litre of the solution

= 1000 mL × 1.84 g mL^{-1} = 1840 g

$$\text{Mass of } H_2SO_4 \text{ in 1 L of solution} = \frac{93 \text{ g} \times 1000 \text{ mL}}{100 \text{ mL}} = 930 \text{ g}$$

So, Mass of water = (1840 – 930 g) = 910 g = 0.91 kg

Thus,

Molality of H_2SO_4

$$= \frac{\text{No. of moles of } H_2SO_4}{\text{Mass of water in kg}} = \frac{930 \text{ g}/98 \text{ g mol}^{-1}}{0.91 \text{ kg}} = 10.43 \text{ mol kg}^{-1}$$

Example 62:

A sugar syrup of weight 214.2 g contains 34.2 g of sugar ($C_{12}H_{22}O_{11}$). Calculate (i) molal concentration, (ii) molefraction of sugar in the syrup.

Solution:

Mass of sugar syrup = 214.2 g

Mass of sugar in the syrup = 34.2 g

So, Mass of water in the syrup = 180.0 g

Molar mass of sugar ($C_{12}H_{22}O_{11}$) = 42 g mol^{-1}

$$\text{Therefore, No. of moles of sugar} = \frac{34.2 \text{ g}}{342 \text{ g/mol}} = 0.1 \text{ mol}$$

So, (i) Molal conc. of sugar

$$= \frac{\text{No. of moles of sugar}}{\text{Mass of solvent in gram}} \times 1000$$

$$= \frac{0.1}{180} \times 1000 \text{ mol kg}^{-1}$$

$$= 0.55 \text{ mol kg}^{-1}$$

(ii) No. of moles of sugar in syrup = 0.1 mol

No. of moles of water in syrup = $\frac{180 \text{ g}}{18 \text{g mol}^{-1}} = 10$ mol

So, Molefraction of sugar

$$= \frac{n_{sugar}}{n_{sugar} + n_{water}} = \frac{0.1}{0.1 + 10} = \frac{0.1}{10.1} = 0.0099 \cong 0.01$$

Example 63:

Concentrated nitric acid used as a laboratory reagent is usually 69% by mass of nitric acid. Calculate the volume of the solution which contained 23 g HCO_3. Density of the conc. HNO_3 solution is 1.41g cm^{-3}.

Solution:

Let, Mass of conc. HCO_3 sample = 100g

So, Mass of HNO_3 in 100 g of sample = 69 g

and Mass of water in 100g of sample = 31 g

Density of conc. HNO_3 sample = 1.41 g cm^{-3}

So, Volume of 100 g of the HNO_3 sample

$$= \frac{\text{Mass}}{\text{Density}} = \frac{100 \text{ g}}{1.41 \text{g cm}^{-3}} = 70.92 \text{cm}^3$$

Thus, 69 g of HNO_3 is contained in 70.92 cm^3 of conc. HNO_3

1g " " $\frac{70.92}{69}$ "

23 g " " $\frac{70.92}{69} \times 23$ " = 23.6 cm^3

Thus, 23.6 cm^3 conc. HNO_3 sample contained 23 g of HNO_3.

Example 64:

The molality and molarity of a solution of sulphuric acid are 4.13 mol/kg and 11.12 mol/L respectively. Calculate density of the solution.

Solution:

Density is related to molarity and molality of a solution by the relationship

$$\rho = \text{M mol L}^{-1}\left(\frac{1}{\text{m mol kg}^{-1}} + \text{M}_{\text{solute}}\left(\text{kg mol}^{-1}\right)\right)$$

$$\text{or } \rho = 11.12 \text{ mol L}^{-1}\left(\frac{1}{94.13 \text{ mol kg}^{-1}} + \frac{98}{1000} \text{ kg mol}^{-1}\right)$$

ρ = 11.12 mol L–1 (0.0106 kg mol^{-1} + 0.098 kg mol^{-1})

or ρ = 11.12 × 0.1086 kg L^{-1} = 1.208 kg L^{-1}

Thus, density of the solution of sulphuric acid is 1.208 kg L^{-1} (or 1.208 g/mL).

Example 65:

Derive the relationship

$$\rho = \text{M}\left(\frac{1}{\text{m}} + \text{M}_{\text{solute}}\right)$$

where ρ *is the density of the solution*

M is the molarity of the solution

m is the molality of the solution

M_{solute} *is the molar mass of the solute in the unit of kg* mol^{-1}.

Solution:

As per definition,

Density of a solution in kg/L

$$= \frac{\text{Mass of the solution in kg}}{\text{Volume of the solution in L}}$$

$$= \frac{\text{Mass of solvent in kg + Mass of solute in kg}}{\text{Volume of the solution in L}}$$

Let m_{solven} be the mass of solvent in the solution in kg

m_{solute} be the mass of solute in the solution in kg

$V_{solution}$ be the volume of the solution

n_{solute} be the number of moles of the solute in the solution

Then, $\rho\ (kg/L) = \dfrac{[m_{solvent}\ (kg) + m_{solute}\ (kg)]}{V_{solution}\ (L)}$

Multiply and divide the RHS by n_{solute} ‾ the unit of n_{solute} is mole.

$$\rho\ (kg/L) = \frac{n_{solute}}{n_{solute}} \times \frac{[m_{solvent}\ (kg) + m_{solute}\ (kg)]}{V_{solution}\ (L)}$$

$$= \frac{n_{solute}}{V_{solution}\ (L)} \times \left[\frac{m_{solvent}\ (kg)}{n_{solute}} + \frac{m_{solute}\ (kg)}{n_{solute}}\right]$$

$$= \text{Molarity}\ (\text{mol L}^{-1}) \times \left[\frac{1}{\text{Molality}\ (\text{mol kg}^{-1})} + m_{solute}\ (\text{kg mol}^{-1})\right]$$

or, $\rho = M\left(\dfrac{1}{m} + M_{solute}\right) kg / L$

Example 66:

The correct expression relating molarity (M), molality (m), density (ρ) and molar mass of solute (M_2) is

(a) $m = \dfrac{M}{\rho + MM_2}$ (b) $m = \dfrac{M}{\rho - MM_2}$

(c) $m = \dfrac{\rho + MM_2}{M}$ (d) $m = \dfrac{\rho - MM_2}{M}$

Solution:

Ans. The correct answer is (b).

$$\text{Molarity, } M = \frac{\text{Amount of solute}}{\text{Volume of solution}} = \frac{n_2}{(n_1M_1 + n_2M_2)/\rho} \quad \text{...(i)}$$

$$\text{Molarity, } m = \frac{\text{Amount of solute}}{\text{Mass of solvent}} = \frac{n_2}{n_1M_1} \quad \text{(ii)}$$

Dividing both the numerator an denominator of RHs of Eq. (i) by n_1, one gets

$$M = \frac{(n_2/n_1)\rho}{M_1 + (n_2/n_1)M_2} \quad ...(iii)$$

substituting the value of n_2/n_1 from Eq. (ii) in Eq. (iii), one gets .

$$M = \frac{mM_1\rho}{M_1 + mM_1M_2} = \frac{m\rho}{1 + mM_2}$$

Rearranging this expression, one gets

$$m = \frac{M}{\rho - MM_2}$$

Example 67:

Write balanced chemical equations for the following :

(i) Sodium nitrite is produced by absorbing the oxides of nitrogen in aqueous solution of washing soda.

(ii) Nitrogen is obtained in the reaction of aqueous ammonia with potassium permanganate.

(iii) Elemental phosphorus reacts with conc. HNO_3 to give phosphoric acid.

(iv) Ethylene glycol is obtained by the reaction of ethylene with potassium permanganate.

(v) Sulphur is precipitated in the reaction of hydrogen sulphide with sodium bisulphite solution.

(vi) Carbon dioxide is passed through a suspension of limestone in water.

Solution:

(i) Skeletal equation : $Na_2CO_3(aq) + NO(g) + NO_2(g) \rightarrow NaNO_2(aq) + CO_2(g)$

Balanced equation : $Na_2CO_3(aq) + NO(g) + NO_2(g) \rightarrow \underset{\text{sodium nitrite}}{2NaNO_2(aq)} + CO_2(g)$

(ii) Skeletal equation : $KMnO_4 + NH_3 \rightarrow MnO_2 + KOH + H_2O + N_2(g)$

Balanced equation : $\underset{\text{pot permanganate}}{2KMnO_4} + 2NH_3 \rightarrow 2MnO_2 + 2KOH + 2H_2O + N_2(g)$

(iii) Skeletal equation : $HNO_3 + P_4 \rightarrow NO_2 + H_3PO_4 + H_2O$

Balanced equation : $20HNO_3 + P_4 \rightarrow 2NO_2 + 4H_3PO_4 + 4H_2O$

(iv) Skeletal equation : $KMnO_4 + C_2H_4 + H_2O \rightarrow MnO_2 + KOH + (CH_2OH)_2$

Balanced equation : $2KMnO_4 + \underset{\text{ethylene}}{3C_2H_4} + 4H_2O \rightarrow 2MnO_2 + 2KOH + \underset{\text{ethylene glycol}}{3(CH_2OH)_2}$

(v) Skeletal equation : $NaHSO_3 + H_2S \rightarrow Na_2SO_3 + H_2O + S$

Balanced equation : $\underset{\text{sodium bisulphite}}{2NaHSO_3} + 2H_2S \rightarrow \underset{\text{sodium sulphite}}{Na_2SO_3} + 3H_2O + 3S$

(vi) Skeletal equation : $\underset{\text{limestone}}{CaCO_3} + CO_2 + H_2O \rightarrow Ca(HCO_3)_2$

Balanced equation : $\underset{\text{(insouble)}}{CaCO_3} + CO_2 + H_2O \rightarrow \underset{\text{(souble)}}{Ca(HCO_3)_2}$

Example 68:

Complete and balance the following chemical reactions :

(i) Copper reacts with HNO_3 to give NO and NO_2 in molar ratio of 2 : 1.

$$Cu + HNO_3 \rightarrow ... + NO + NO_2 + ...$$

(ii) Na2CO_3 is added to a solution of copper sulphate.

$$CuSO_4 + Na_2CO_3 + H_2O \rightarrow ... + Na_2SO_4 + ...$$

(iii) Red phosphorus is reacted with iodine in presence of water .

$$P + I_2 + H_2O \rightarrow ... + ...$$

(iv) Anhydrous potassium nitrate is heated with excess of metallic potassium.

$$KNO_3(s) + K(s) \rightarrow ... + ...$$

(v) Potassium dichromate and concentrated hydrochloric acid are heated together.

$$K_2Cr_2O_7 + HCl \rightarrow KCl + ... + ... + H_2O$$

Solution:

(i) Copper with dil. HNO_3 gives NO whereas with HNO_3, it gives NO_2. The individual reactions are

$$3Cu + 8HNO_3 \text{ (dil)} \rightarrow 2NO + 3Cu(NO_3)_2 + 4H_2O$$

$$Cu + 4HNO_3 \text{ (conc.)} \rightarrow 2NO_2 + Cu(NO_3)_2 + 2H_2O$$

To get molar ration of 2 : 1 for NO and NO_2, multiply former equation by 2 and add it to the latter,

$$7Cu + 20HNO_3 \rightarrow 7Cu(NO_3)_2 + 4NO + 2NO_2 + 10H_2O$$

(ii) When Na_2CO_3 is added to copper sulphate solution, $Cu(OH)_2$ is precipitated.

$$CuSO_4 + Na_2CO_3 + H_2O \rightarrow Cu(OH)_2 + Na_2SO_4 + CO_2$$

(iii) The reaction of phosphorus with iodine in the presence of water is

$$2P + 3I_2 + 6H_2O \rightarrow 2H_3PO_3 + 6HI$$

(iv) When anhydrous potassium nitrate is heated with excess of metallic potassium, potassium oxide is formed.

$$2KNO_3(s) + 10K(s) \rightarrow 6K_2O(s) + N_2$$

(v) When potassium dichromate and conc. HCl are heated together, HCl gets oxidized to chlorine.

$$K_2Cr_2O_7 + 8HCl \rightarrow 2KCl + 2CrCl_3 + 4H_2O + 3O$$

$$2HCl + O \rightarrow [H_2O + Cl_2] \times 3$$

$$K_2Cr_2O_7 + 14HCl \rightarrow 2KCl + 2CrCl_3 + 7H_2O + 3Cl_2$$

Example 69:

(i) Complete and balance the following :

(a) $NH_3 + Na0Cl \rightarrow ... + ... + ...$

(b) $AgBr + Na_2S_2O_3 \rightarrow ... + ...$

(c) $(NH_4)2S_2O_8 + H_2O + MnSO_4 \rightarrow ... + ... + ...$

(ii) The acidic aqueous solution of ferrous ions forms a brown complex in the presence of NO_3^-, by the following two steps. Complete and balance the equations.

$$[Fe(H_2O)_6]^{2+} + NO_3^- + H^+ \rightarrow ... [Fe(H_2O)_6]^{3+} + H_2O$$

$$[Fe(H_2O)_6]^{2+} + ... \rightarrow ... + H_2O$$

(iii) Give briefly the process of isolation of magnesium from sea water by Dow's process. Give equations for the steps involved.

Solution:

(i)(a) $2NH_3 + NaOCl \rightarrow NH_2NH_2 + NaCl + H_2O$

(b) $AgBr + 2Na_2S_2O_3 \rightarrow Na_3\ [Ag(S_2O_3)_2] + NaBr$

(c) $(NH_4)_2S_2O_8 + 2H_2O + MnSO_4 \rightarrow (NH_4)_2SO_4 + MnO_2 + 2H_2SO_4$

(ii) $3[Fe(H_2O)_6]^{2+} + NO_3^- + 4H^+ \rightarrow NO + 3[Fe(H_2O)_6]^{3+} + 2H_2O$

$$[Fe(H_2O)_6]^{2+} + NO \rightarrow [Fe(H_2O)_5NO]^{2+} + H_2O$$

(iii) In Dow's process, magnesium is recovered from sea water through the following steps: Slaked lime ($Ca(OH)_2$) is added to sea water to precipitate magnesium hydroxide which is subsequently dissolved in hydrochloric acid.

$$\underset{\text{In sea water}}{Mg^{2+}} + Ca(OH)_2 \rightarrow Mg(OH)_2(s) + Ca^{2+}$$

$$Mg(OH)_2 + 2HCl \rightarrow \begin{matrix} MgCl_2 \\ \downarrow \text{crystallisation} \\ MgCl_2.6H_2O \end{matrix} + 2H_2O$$

Hydrated magnesium chloride, $MgCl_2.6H_2O$, thus obtained is partially dehydrated by heating in a current of dry hydrogen chloride gas. The magnesium chloride so obtained is mixed with a fused mixture of sodium chloride and calcium chloride.

Magnesium chloride melts in the fused mixture (975 – 1025 K) and becomes completely anhydrous. The molten mixture of chlorides is then electrolysed and magnesium is obtained at the cathode and chlorine is liberated at the anode

$$MgCl^2 \rightarrow Mg^{2+} + 2Cl^-$$

At cathode : $Mg^{2+} + 2e^- \rightarrow Mg$

At anode : $2Cl^- \rightarrow Cl_2 + 2e^-$

Example 70:

Complete and balance the following equation :

(i) $[MnO_4]^{2-} + H^+ \rightarrow ... + [MnO_4]^- + H_2O$

(ii) $Ca_5(PO_4)_3F + H_2SO_4 + H_2O \xrightarrow{\text{heat}} ...\ 5CaSO_4\ 2H_2O + ...$

(iii) $SO_2\ (aq) + Cr_2O_7^{2-} + 2H+ \rightarrow ... + ... + ...$

(iv) $Sn + 2KOH + 4H_2O \rightarrow ... + ...$

Solution:

The balanced equation are given below :

(i) $3\ MnO_4^- + 4H^+ \rightarrow MnO_2 + 2MnO_4^- + 2H_2O$

(ii) $Ca_5(PO_4)_3 F + 5H_2SO_4 + 10H_2O \xrightarrow{heat} 3H_3PO_4 + 5CaSO_4\ 2H_2O + HF$

(iii) $3SO_2(aq) + Cr_2O_7^{2-} + 2H^+ \rightarrow 2Cr^{3+} + 3SO_4^{2-} + H_2O$

(iv) $Sn + 2KOH + 4H_2O \rightarrow K_2Sn(OH)_6 + 2H_2$

Example 71:

4 litres of water are added to 2 L of 6 molar hydrochloric acid solution. What is the molarity of the resulting solution ?

Solution:

Initial volume, $V_1 = 2$ L   Final volume, $V_2 = 4$ L + 2 L = 6 L

Initial molarity, M1 = 6 M   Final molarity = ? (say M_2)

Then $M_1 V_1 = M_2 V_2$

$6\ M \times 2\ L = M_2 \times 6\ L$

So, $$M_2 = \frac{6M \times 2L}{6L} = 2\ M$$

Thus, the resulting solution is 2 M solution of hydrochloric acid.

Example 72:

What volume of 10 M HCl and 3 M HCl should be mixed to obtain 1 L of 6 M HCl solution ?

Solution:

Let, the required volume of 10 M HCl be V litre. Then, the required volume of 3 M HCl be (1 – V) litre. Using molarity equations,

$$M_1 V_1 + M_2 V_2 = M_3 V_3$$

$$10 \times V + 3 \times (1 - V) = 6 \times 1$$

$$10\ V + 3 - 3\ V = 6$$

$$7\ V = 6 - 3 = 3$$

$$3/7\ L = 0.428\ L = 428\ mL$$

Thus, Volume of 10 M HCl required = 428 mL

and Volume of 3 M HCl required = (1000 mL–428 mL) = 572mL

Example 73:

Calculate number of atoms of each type in 5.3 g of Na_2CO_3.

Solution:

Molar mass of Na_2CO_3 = (2×23.0) + (1×12.0) + (3×16.0)g mol^{-1}

= (46.0+12.0+48.0)g mol^{-1} =106.0g mol^{-1}

So,

Number of moles of Na_2CO_3 in 5.3 g of Na_2CO_3

$$= \frac{5.3 \text{ g}}{106.0 \text{ g mol}^{-1}} = 0.05 \text{ mol}$$

Now,

Na_2CO_3 ≡	2 Na	+	C	+	3 O
1 mol	2 mol		1 mol		3 mol
0.05mol	2 × 0.05 mol		1 × 0.05 mol		3 × 0.05 mol
	= 0.1 mol		= 0.05 mol		= 0.15 mol

Since, one mole of any substance contains 6.02×10^{23} chemical units, hence

No. of Na atoms in 5.3 g of Na_2CO_3 = 0.1 mol × 6.023×10^{23} mol^{-1}

= 6.023×10^{22}

No. of C atoms in 5.3 g of Na_2CO_3 = 0.05 mol × 6.023×10^{23} mol^{-1}

= 3.011×10^{22}

No. of O atoms in 5.3 g of Na_2CO_3 = 0.15 mol × 6.023×10^{23} mol^{-1}

= 9.034×10^{22}

Example 74:

How many years would it take to spend Avogadro number of rupees at the rate of Rs. 10 lakh per second ?

Solution:

We know,

Avogadro's number $= 6.023 \times 10^{23}$

So, Total money to be spent $=$ Rs. 6.023×10^{23}

Time taken to spend Rs. 10 lakh ($= 10^6$) $= 1$ s

Therefore,

Time taken to spend s. $6.023 \times 10^{23} = \dfrac{1\text{s}}{10^6} \times 6.023 \times 10^{23}$

$= 6.023 \times 10^{17}$ s

The time in seconds may be converted into years as follows.

$$6.023 \times 10^{17} \text{ s} = \frac{6.023 \times 10^{17}}{60 \times 60 \times 24 \times 365} \text{ yr} = 1.91 \times 10^{10} \text{ yr}$$

So, the time taken to spend Avogadro's number of rupees is 1.91 10^{10} years.

Example 75:

Which is cheaper? 40% HCl at the rate of Rs 6 per kg, or 80% H_2SO_4 at the rate of Rs 2.5 per kg required to neutralise 7 kg of KOH.

Solution:

The neutralisation reaction using HCl is

KOH	+	HCl	→	KCl	+	H_2O
(39+16 + 1) g		(1 + 35.5) g				
56 g		36.5 g				
56 g		63.5 kg				
7 kg		$\dfrac{36.5 \text{ kg}}{56 \text{ kg}} \times 7 \text{ kg} = 4.56$ kg				

Mass of 40% HCl required to neutralise 7 kg of KOH = 4.56 kg $\times \dfrac{100}{40} = 11.4$ kg

Total cost of using HCl = Rs 6.00/ kg × 11.4 kg = Rs. 68.40

The neutralisation reaction using H_2SO_4 is

KOH	+	H2SO_4	→	K_2SO_4 +	$2H_2O$
2×(39+16+1)g		(2×1+32+4×16) g			
112 g		98 g			

or 112 kg 98 kg

7 kg $\frac{98\ kg}{112\ kg}\times 7kg = 6.12\ kg$

Mass of 80% H_2SO_4 required to neutralise 7 kg of KOH

Total cost of using H_2SO_4 = Rs 2.5/kg × 7.65 kg = Rs. 19.12

Thus, the use of H_2SO_4 in neutralising 7 kg KOH costs less. Therefore, the use of 80% H_2SO_4 is cheaper.

Example 76:

Calculate the total number of electrons present in 1.6 g of methane.

Solution:

The molecular formula of methane is CH_4. Then

Molar mass of methane = (12 + 4 × 1) g/mol = 16 g/mol

Mass of methane = 1.6 g

So,

No. of molecules in 1.6 g of methane = $\frac{1.6\ g}{16\ g/mol}\times 6.02\times 10^{23}/mol$

$= 6.02 \times 10^{22}$

From the structure,

No. of electrons in a methane molecule = 6 + 4 =10

So, No of electrons in 1.6 g of methane = $10 \times 6.02 \times 10^{22}$

$= 6.02 \times 10^{23}$.

Example 77:

1.0 g of an alloy of aluminium and magnesium when treated with excess of dill . HCl form magnesium chloride, aluminium chloride and hydrogen gas collected over mercury at 0°C has a volume of 1.20 L at 0.92 atm pressure. Calculate the composition of the alloy.

Solution:

This problem is solved through the following steps.

Step 1 : Calculation of volume of hydrogen at NTP

$P_1 = 0.92$ atm $P_2 = 1$ atm

$V_1 = 1.20$ L $\qquad V_2 = ?$

$T_1 = 0°C = 273$ k $\qquad T_2 = 273$ k

Using the relationship, $\frac{P_1V_1}{T_1} = \frac{P_2V_2}{T_2}$

one gets $V_2 = \frac{P_1V_1}{T_1} \times \frac{T_2}{P_2} = \frac{0.92 \text{ atm} \times 1.20\text{L} \times 273 \text{ K}}{273\text{k} \times 1 \text{ atm}}$

Step 2: Calculating the volume of hydrogen evolved during the reaction

Total mass of the alloy = 1.0 g

Let, Mass of aluminium = x g

So, Mass of magnesium = (1.0 – x)g

Aluminium and magnesium react with dil. HCl as follows.

$$2\,Al + 6\,HCl \rightarrow 2\,AlCl_3 + 3\,H_2$$

2 × 27 g $\qquad$ 3 × 22.4 L

54 g $\qquad$ 67.2 L

x g $\qquad \frac{67.2\text{L}}{54\text{ g}} \times \text{x g} = \frac{67.2\text{x}}{54}\text{ L}$

and $$Mg + 2\,HCl \rightarrow MgCl_2 + H_2$$

24 g $\qquad$ 22.4 L

(1.0 – x)g $\qquad \frac{22.4\text{ L}}{24\text{g}} \times (1.0 - \text{x})\text{g} = \frac{22.4\,(1.0 - \text{x})}{24}\text{ L}$

Thus,

Total volume of hydrogen liberated at NTP

$$= \frac{67.2\text{x}}{54}\text{ L} + \frac{22.4\,(1.0 - \text{x})}{24}\text{ L}$$

This volume must be equal to 1.10 L. So

Step 3: Solving for x

Equating the expressions for volume

$$\frac{67.2\text{x}}{54} + \frac{22.4\,(1.0 - \text{x})}{24} = 1.10$$

This gives, $x = 0.5486$

Therefore,

Mass of aluminium in 1.0 g of alloy = 0.5486 g

and Percentage of aluminium in the alloy = $\frac{0.5486\text{ g}}{1.0\text{ g}} \times 100 = 54.86$

Then, Percentage of magnesium in the alloy = (100 – 54.86) = 45.14

Therefore, the alloy contains 54.60% aluminium and 45.14% magnesium.

Example 78:

1.0 g magnesium is burnt in a closed container which contains 0.6 g of oxygen.

(a) Which reactant is left in excess?

(b) Find the mass of the excess reactant.

(c) How many milliliters of 0.5 N H_2SO_4 will dissolved the resides in the vessel ?

Solution:

The reaction is

2 Mg	+	O_2	→	2 MgO
2 × 24 g		2 × 16 g		2 × (24 + 16) g
48 g		32 g		80 g
1 g		$\frac{32\text{ g}}{48\text{g}} \times 1\text{ g} = 0.667\text{ g}$		

(a) Thus, 1 g of magnesium would require 0.667 g of oxygen for complete combustion. In the container only 0.6 g of oxygen is available. Therefore, a part of magnesium is left unreacted.

(b) From the reaction stoichiometry,

32 g oxygen react with 48 g of Mg

0.6 g " " " $\frac{48\text{ g}}{32\text{g}} \times 0.6\text{ g} = 0.9\text{ g}$

Therefore,

Mass of excess magnesium left behind = 1.0 g – 0.9 g = 0.1 g

Thus, the residue will contain both Mg and MgO. Both Mg and MgO dissolve in H2SO4 due to the following reactions,

$$Mg + H_2SO_4 \rightarrow MgSO_4 + H_2$$

$$MgO + H_2SO_4 \rightarrow MgSO_4 + H_2O$$

From these reactions, H_2SO_4 equivalent to 1 g of Mg (Total magnesium = Unreacted Mg + Mg reacted to form MgO) is needed to dissolved the residue. Therefore,

1 mol of Mg requires 1 mol of H_2SO_4

24 g of Mg requires 1 mol of H_2SO_4

So, 1 g of mg requires

$$\frac{1g}{24\,g} \times 1 \text{ mol of } H_2SO_4 = \frac{1}{24} \text{mol of } H_2SO_2 \quad ...(a)$$

Let, V mL of 0.5 N H_2SO_4 be used for complete dissolution of the residue, then

No. of moles of H_2SO_4 in V mL of 0.5 N H_2SO_4

$$= 1/2 \left(N_{H_2SO_2} \times V_{H_2SO_4}\right)/1000$$

$$= 1/2 \times 0.5 \times V/1000 \quad ...(b)$$

From Eqs (a) and (b),

$$\frac{1}{2} \times \frac{0.5\,V}{1000} = \frac{1}{24}$$

or $$V = \frac{2 \times 1000}{24 \times 0.5} = 166.7 \text{ mL}$$

Thus, 166.7 mL of 0.5 N H_2SO_4 is needed to dissolve the residue.

Example 79:

8.0575 × 10^{-2} kg of glauber's salt is dissolved in water a obtain 1 dm^3 of a solution of density 1077.2 kg m^{-3}. Calculate the molarity, molality and molefraction of Na_2SO_4 in the solution.

Solution:

Glauber's salt is Na_2SO_4 $10H_2O$. Therefore,

Molar mass of Glauber's salt = {2 × 23.00 + 32.10 + 4 × 16.00 + 10 (2 × 1.01 + 16.00)} g mol^{-1}

= 322.3 g mol^{-1}

Amount of Glauber's salt dissolved,

$$n_2 = \frac{m}{M} = \frac{8.0575 \times 10^{-2}\ \text{kg}}{322.3 \times 10^{-3}\ \text{kg mol}^{-1}} = 0.25\,\text{mol}$$

Mass of dm^3 of solution = Volume × Density

= (10^{-3} m^3) × (1077.2 kg m^{-3})

= 1.0772 kg

Mass of Na_2SO_4 in 1 dm^3 solution = No. of moles of Na_2SO_4 (n_2) × Molar mass of Na_2SO_4

= (0.25 mol) (2 × 23 + 32.10 + 4 × 16.00) g mol^{-1}

= 3.5525 × 10^{-2} kg

Mass of solvent in 1 dm^3 solution = (1.0772 – 3.5525 × 10^{-2}) kg

= 1.041675 kg

So, Molarity of solution $= \frac{n_2}{V} = \frac{0.25\ \text{mol}}{1\ \text{dm}^3} = 0.25\ \text{mol dm}^{-3}$

Molality f solution $= \frac{n_2}{m_{solvent}} = \frac{0.25\ \text{mol}}{1.0417\ \text{kg}} = 0.24\ \text{mol kg}^{-1}$

Amount of solvent in 1 dm^{-3} solution,

$$n_1 = \frac{1.0417 \times 10^3\ \text{g}}{18\ \text{g mol}^{-1}} = 57.87\ \text{mol}$$

So,

Molefraction of Na_2SO_4 in solution

$$= \frac{n_1}{n_1 + n_2} = \frac{0.25}{0.25 + 57.87} = 0.0043$$

Example 80:

A crystalline compound when heated becomes anhydrous by losing 51.2% of the mass. The analysis of the anhydrous salt gave the following results, Mg = 20.0%, S = 26.66% and O = 53.33%.

Solution:

(a) Determination of the empirical formula of the anhydrous salt

Element	*Mass %*	*Atomic mass*	*Atomic ratio* $= \frac{\text{Mass \%}}{\text{Atomic mass}}$	*Simplest ratio*	*Simplest whole number ratio*
Mg	20.0	24	20/24 = 0.83	0.83/0.83 = 1	1
S	26.66	32	26.66/32 = 0.83	0.83/0.83 = 1	1
O	53.33	16	53.33/16 = 3.33	3.33/0.83 = 4.01	4

Therefore, the empirical formula of the compound is $MgSO_4$.

(b) Determination of the molecular formula of the anhydrous salt

From the formula,

Empirical formula mass = 24 u + 32 u + (4 × 16 u) = 120 u

Molecular mass (given) = 120 u

So, $$n = \frac{\text{Molecular mass}}{\text{Empirical formula mass}} = \frac{120\,u}{120\,u} = 1$$

So, $$\text{Molecular formula} = 1 \times \text{Empirical formula} = 1 \times MgSO_4 = MgSO_4$$

(c) Determination of the molecular formula of the crystalline salt

Let the molecular formula of the cyrstalline salt be $MgSO_4.xH_2O$

So, one mole of the anhydrous salt contains x moles of water. Then,

Mass of $MgSO_4$ = 120 g

Mass of water = x × 18 g = 18x g

Therefore, Total mass = (120 + 18x) g

Then, $$\%\text{ of water} = \frac{18\,x}{(120 + 18x)} \times 100$$

From the given data, percentage of water is 51.2% . Then,

$$51.2 = \frac{18\,x \times 100}{(120 + 18x)}$$

This gives $$x = \frac{120 \times 51.2}{48.8 \times 18} = 6.99 = 7$$

(rounded to the nearest whole number)

So, the molecular formula of the crystalline salt is $MgSO_4 . 7H_2O$.

Example 81:

Rewrite the following equation in the blanched form showing in it that Al $(OH)_3$ is an insoluble product.

$$Al_2(SO_4)_3 + NaOH \rightarrow Al(OH)_3 + Na_2SO_4$$

Solution:

(i) The inspection of the equation,

$$Al_2(SO_4)_3 + NaOH \rightarrow Al(OH)_3 + Na_2SO_4$$

Shows that there is no elementary gas involved in this reaction. So, balancing is started from the molecule $Al_2(SO_4)_3$ (containing the maximum number of atoms).

(ii) There are 2Al atoms on the left and only 1 on the right. So, Al atoms can be balanced by multiplying $Al(OH)_3$ (on the right) by 2, *i.e.*,

$$Al_2(SO_4)_3 + NaOH \rightarrow 2Al(OH)_3 + Na_2SO_4$$

(iii) Now, there are six OH groups on the right, and only 1 on the left. The – OH groups can be balanced by multiplying NaOH by 6. So, the chemical equation can be written as,

$$Al_2(SO_4)_3 + 6NaOH \rightarrow 2Al(OH)_3 + Na_2SO_4$$

(iv) Now, Al atoms and OH groups are balance. To balanced Na atoms and SO_4 groups multiply Na_2SO_4 on the right by 3. The chemical equation now become s

$$Al_2(SO_4)_3 + 6NaOH \rightarrow 2Al(OH)_3 + 3Na_2SO_4$$

This is the balanced equation. $Al(OH)_3$ is an insoluble product. This information is included in the equation by adding either (s) or an arrow (↓) pointing downward immediately after $Al(OH)_3$. So, the balanced chemical equation, showing that $Al(OH)_3$ is an insoluble product is,

$$Al_2(SO_4)_3 + 6NaOH \rightarrow 2Al(OH)_3\ (s) + 3Na_2SO_4$$

or $$Al_2(SO_4)_3 + 6NaOH \rightarrow 2Al(OH)_3\ (\downarrow) + 3Na_2SO_4$$

Example 82:

Balance the following equation :

(i) $H_3PO_2 \rightarrow H_3PO_4 + PH_3$

(ii) $Ca + H_2O \rightarrow Ca(OH)_2 + H_2$

(iii) $Fe(SO_4)_3 + NH_3 + H_2O \rightarrow Fe(OH)_3 + (NH_4)_2SO_4$

Solution:

(i) The given equation : $H_3PO_2 \rightarrow H_3PO_4 + PH_3$

Balancing of P on both the sides requires a coefficient of 2 before H_3PO_2

Thus, $2H_3PO_2 \rightarrow H_3PO_4 + PH_3$

(ii) $Ca + H_2O \rightarrow Ca(OH)_2 + H_2$

First, balance Ca atoms (already balanced)

Then, balance H atoms–this requires multiplication of H_2O by 2, hence

$$Ca + 2H_2O \rightarrow Ca(OH)_2 + H_2$$

The O atoms get balanced in the process. Therefore, balanced equation is

$$Ca + 2H_2O \rightarrow Ca(OH)_2 + H_2$$

(iii) $Fe_2(SO_4)_3 + NH_3 + H_2O \rightarrow Fe(OH)_3 + (NH_4)_2SO_4$

(a) Balance Fe atoms – it requires multiplication of $Fe(OH)_3$ by 2. Thus, the partly balanced equation is

$$Fe_2(SO_4)_3 + NH_3 + H_2O \rightarrow 2Fe(OH)_3 + (NH_4)_2SO_4$$

(b) Balance SO_4^{2-} ion – it require 3 before $(NH_4)_2SO_4$. The resulting equation is

$$Fe_2(SO_4)_3 + NH_3 + H_2O \rightarrow 2Fe(OH)_3 + 3(NH_4)_2SO_4$$

(c) To balance N atoms multiply NH_3 by 6. The resulting equation is,

$$Fe_2(SO_4)_3 + 6NH_3 + H_2O \rightarrow 2Fe(OH)_3 + 3(NH_4)_2SO_4$$

(d) H an d O atoms can be balanced by multiplying H_2O by 6. The completely balanced equation is,

$$Fe_2(SO_4)_3 + 6NH_3 + 6H_2O \rightarrow 2Fe(OH)_3 + 3(NH_4)_2SO_4$$

Example 83:

Balance the equation, $NaOH + Cl_2 \rightarrow NaCl + NaClO_3 + H_2O$ *by partial equation method.*

Solution:

(i) Skeleton equation, $NaOH + Cl_2 \rightarrow NaCl + NaClO_3 + H_2O$, can be spit into two partial equations.

Partial eq. 1 : $NaOH + Cl_2 \rightarrow NaCl + NaClO + H_2O$

Partial eq. 2 : $NaClO \rightarrow NaClO_3 + NaCl$

(ii) Balancing the two partial equations by hit and trial method gives,

Balanced partial eq. 1 : $2NaOH + Cl_2 \rightarrow NaCl + NaClO + H_2O$

Balanced Partial eq. 2: $3\ NaClO \rightarrow NaClO_3 + 2\ NaClO$

(iii) NaClO does not appear in the equation for the overall reaction. So, to exactly cancel it, the balanced partial equation 1 is multiplied by 3. Then the matched, balanced partial equations and the balanced over al equation are,

$$2\ NaOH + Cl_2 \rightarrow NaCl + NaClO + H_2O\ ] \times 3$$

$$3\ NaClO \rightarrow NaClO_3 + 2\ NaCl$$

$$6\ NaOH + 3Cl_2 \rightarrow 5NaCl + NaClO_3 + 3H_2O$$

(Balanced chemical equation)

Example 84:

$KMnO_4$ in an acidic solution, oxidizes ferrous sulphate to ferric sulphate. Write the skeleton equation for this reaction and balance it by partial equation method .

Solution:

(i) The skeleton equation is,

$$KMnO_4 + H_2SO_4 + FeSO_4 \rightarrow K_2SO_4 + MnSO_4 + Fe_2(SO_4)_3 + H_2O$$

(ii) This reaction is assumed to proceed through the following reactions .

Partial equation 1 : $KMnO_4 + H_2SO_4 \rightarrow K_2SO_4 + MnSO_4 + H_2O + O$

Partial equation 2 : $FeSO_4 + H_2SO_4 + O \rightarrow Fe_2(SO_4)_3 + H_2O$

Following the principle of hit and trial method, the two equations given above are balanced and are suitably multiplied to cancel out the common element, as shown below :

Balanced Partial eq 1 : $2KMnO_4 + 3H_2SO_4 \rightarrow K_2SO_4 + 2MnSO_4 + 3H_2O + 5O$

Balanced Partial eq 2 : $2FeSO_4 + H_2SO_4 + O \rightarrow Fe_2(SO_4)_3 + H_2O] \times 5$

Balanced over-

equation $2KMnO_4 + 8H_2SO_4 + 10FeSO_4 \rightarrow K_2SO_4 + 2MnSO_4 + 5Fe_2(SO_4)_3 + 8H_2O$

Example 85:

Write the skeleton equation for the reaction between acidified potassium permanganate and oxalic acid. Balance the skeleton equation by the partial equation method.

Solution:

The reactants and products in this reaction are,

Reactants : Potassium permanganate ($KMnO_4$), Sulphuric acid (H_2SO_4), Oxalic acid ($H_2C_2O_4$).

Products : Potassium sulphate (K_2SO_4), Manganese sulphate ($MnSO_4$), Water (H_2O) and Carbon dioxide (CO_2).

Then, the skeleton equation is,

$$KMnO_4 + H_2SO_4 + H_2C_2O_4 \rightarrow K_2SO_4 + MnSO_4 + H_2O + CO_2$$

(i) The overall reaction is split into the two reactions represented by the following partial equations.

Partial eq 1 : $KMnO_4 + H_2SO_4 \rightarrow K_2SO_4 + MnSO_4 + H_2O + O$

Partial eq 2 : $H_2C_2O_4 + O \rightarrow 2CO_2 + H_2O$

(ii) These partial equations are balanced by hit and trial method as illustrated above. The balanced partial equations are,

Balanced Partial eq 1 : $2KMnO_4 + 3H_2SO_4 \rightarrow K_2SO_4 + 2MnSO_4 + 3H_2O + 5O$

Balanced Partial eq 2 : $H_2C_2O_4 + O \rightarrow 2CO_2 + H_2O$

(iii) Balancing the above balanced partial equations with respect to each other, *i.e.*, by balancing the number of O atoms evolved and consumed, one gets,

$$2\ KMnO_4 + 3H_2SO_4 \rightarrow K_2SO_4 + 2MnSO_4 + 3H_2O + 5O \quad ...(a)$$

$$H_2C_2O_4 + O \rightarrow 2CO_2 + H_2O\ ] \times 5 \quad ...(b)$$

(iv) The balanced chemical equation for the overall reaction is obtained by adding the above equations (Eqs. (a) and (b).

$$2\ KMnO_4 + 3H_2SO_4 \rightarrow K_2SO_4 + 2MnSO_4 + 3H_2O + 5O$$

$$H_2C_2O_4 + O \rightarrow 2CO_2 + H_2O\ ] \times 5$$

$$2KMnO_4 + 3H_2SO_4 + 5H_2C_2O_4 \rightarrow K_2SO_4 + 2MnSO_4 + 8H_2O + 10CO_2$$

Example 86:

Balance the chemical equation,

$$KMnO_4 + HCl \rightarrow KCl + MnCl_2 + H_2O + Cl_2$$

by hit and trial method.

Solution:

(i) Skeleton equation : $KMnO_4 + HCl \rightarrow KCl + MnCl_2 + H_2O + Cl_2$

(ii) Writing elementary gas in atomic form : $KMnO_4 + HCl$

$\rightarrow KCl + MnCl_2 + H_2O + Cl$

(iii) Starting from $KMnO_4$: K and Mn are balanced.

(iv) Balancing O atoms : $KMnO_4 + HCl$

$\rightarrow KCl + MnCl_2 + 4H_2O + Cl$

(v) Balancing H atoms : $KMnO_4 + 8HCl$

$\rightarrow KCl + MnCl_2 + 4H_2O + Cl$

(vi) Balancing Cl atoms : $KMnO_4 + 8HCl$

$\rightarrow KCl + MnCl_2 + 4H_2O + 5Cl$

This is balanced atomic equation.

(vii) Making it molecular : $2KMnO_4 + 16HCl$

$\rightarrow 2KCl + 2MnCl_2 + 8H_2O + 5Cl_2$

So, the balanced molecular equation is,

$2KMnO_4(aq) + 16HCl(aq)$

$\rightarrow 2KCl(aq) + 2MnCl_2(aq) + 8H_2O(l) + 5Cl_2(g)$

EXERCISES

1. Define chemical equation.
2. Name the law which requires balancing of the chemical equations.
3. Why is it essential to balance a chemical equation ?
4. What condition must be satisfied by a balanced chemical equation ?
5. Define the limiting reagent.
6. State the law which must hold good for a balanced chemical equation.
7. Is the following reaction exothermic or endothermic ? Give reason.

$$C_2H_5OH(l) + 3O_2(g) \rightarrow 2\ CO_2(g) + 3H_2O(l) + 1368\ kJ$$

8. Write the balanced chemical equation for the reaction between calcium carbonate and hydrochloric acid to form calcium chloride, carbon dioxide and water.
9. Rewrite the following in formation in the form of a balanced chemical equation: "magnesium displaces hydrogen form dilute sulphuric acid."
10. Rewrite the following chemical equation in balanced form showing in it that $Al(OH)_3$ is an insoluble product.

$$Al_2(SO_4)_3 + NaOH \rightarrow Al(OH)_3 + Na_2SO_4$$

11. Translate the following statements into balanced chemical equations :

    (a) when barium chloride solution is mixed which zinc sulphate solution, zinc chloride and a precipitate or barium sulphate are formed.

    (b) aluminium metal displaces iron from iron oxide (Fe_2O_3), giving aluminium oxide and iron.

    (c) manganese dioxide reacts with hydrochloric acid to give manganous chloride, water and chlorine gas.

    (d) sulphur dioxide reacts with hydrogen sulphide to form sulphur and water.

    (e) methane burns in air to give carbon dioxide, water and heat is evolved.

(f) magnesium burns in an atmosphere of carbon dioxide to form magnesium oxide and carbon.

12. Balance the following chemical equations by hit and trial method.

(a) $MgCl_2(aq) + NaOH\ (aq) \rightarrow Mg(OH)_2\ (s) + NaCl\ (aq)$

(b) $BaCO_2(s) + HNO_3\ (aq) \rightarrow Ba(NO_3)_2\ (aq) + CO_2\ (g) + H_2O(l)$

(c) $Ba_2SO_3 + H_2O + Cl_2 \rightarrow Na_2SO_4 + HCl$

(d) $HCl + O_2 \rightarrow H_2O + Cl_2$

(e) $N_2H_4 + H_2O_2 \rightarrow N_2 + 4H_2O$

(f) $KClO_3 \xrightarrow{heat} KCl\ (s) + 3O_2(g)$

(g) $Al_2O_3 + NaOH \rightarrow NaAlO_2 + H_2O$

(h) $Fe_3O_4 + HCl \rightarrow FeCl_3 + FeCl_2 + H_2O$

(i) $Cu(NO_3)_2 \rightarrow CuO + NO_2 + H_2O$

(j) $HNO_3 + H_2S \rightarrow NO_2 + H_2O + S$

(k) $Cu + H_2SO_4 \rightarrow CuSO_4 + H_2O + SO_2$

(l) $NH_4Cl + Ca(OH)_2 \rightarrow CaCl_2 + NH_2 + H_2O$

(m) $NO_2 + NaOH \rightarrow NaNO_2 + NaNO_3 + H_2O$

(n) $P_4 + HNO_3 \rightarrow H_3PO_4 + NO_2 + H_2O$

(o) $P_4 + H_2SO_4 \rightarrow H_3PO_4 + SO_2 + H_2O$

(p) $Ca_3(PO_4)_2 + H_3PO_4 \rightarrow Ca(H_2PO_4)_2$

(q) $Fe + H_2O(steam) \rightarrow Fe_3O_4 + H_2\ (g)$

13. Balance the following equations by partial equation method.

(a) $FeSO_4 + H_2SO_4 + K_2Cr_2O_7 \rightarrow Fe_2(SO_4)_3 + Cr_2(SO_4)_3 + K_2SO_4 + H_2O$

(b) $FeSO_4 + H_2SO_4 + O_3 \rightarrow Fe_2(SO_4)_3 + H_2O$

(c) $HNO_3 + C \rightarrow H_2CO_3 + NO_2 + H_2O$

(d) $KI + O_3 + H_2O \rightarrow KOH + O_2 + I_2$

(e) $FeSO_4 + H_2SO_4 + HNO_2 \rightarrow Fe_2(SO_4)_3 + NO + H_2O$

(f) $FeSO_4 + H_2SO_4 + HNO_3 \rightarrow Fe_2(SO_4)_3 + H_2O + NO_2$

(g) $H_2C_2O_4 + H_2SO_4 + KMnO_4 \rightarrow K_2SO_4 + MnSO_4 + CO_2 + H_2O$

(h) $H_2C_2O_4 + C_2H_5OH \rightarrow (C_2H_5COO)_2 + H_2O$

(i) $HNO_3 + As \rightarrow H_3AsO_4 + NO_2 + H_2O$

(j) $Zn + HNO_3\ (dil) \rightarrow Zn(NO_3)_2 + N_2O + H_2O$

(k) $Zn + HNO_3\ (v.\ dilute) \rightarrow Zn(NO_3)_2 + NH_4NO_3 + H_2O$

(l) Zn	+ HNO_3 (conc).	→ $Zn(NO_3)_2$	+ NO_2	+ H_2O
(m) Cu	+ conc. HNO_3	→ $Cu(NO_3)_2$	+ H_2O	+ NO_2
(n) Cu	+ dil. HNO_3	→ $Cu(NO_3)_2$	+ H_2O	+ NO

14. Identify the limiting reagent in the following reactions :
    (a) 6.5 g of zinc added into an aqueous solution containing 12.0 g of sulphuric acid produces 2.24 L of hydrogen gas
    (b) A solution containing 7.4 g of $Ca(OH)_2$ is treated with a sample of hard water containing 15.0 g of $Ca(HCO_3)_2$.
    (c) 5.6 g of iron is heated in the presence of excess of oxygen.
15. An hourly energy requirement of an astronaut can be satisfied by the energy released when 34 g of sucrose ($C_{12}H_{22}O_{11}$) are burnt in his body. How many grams of oxygen would be need to be carried in space capsule to meet his requirement for one day ?
16. A chemist wishes to prepare 6.023×10^{24} molecules of SO_2 according to the reaction,

$$S + O_2 \rightarrow SO_2$$

How many grams of sulphur and oxygen and needed for it ?

17. Sulphuric acid is manufactured by a process represented by the reaction:

$$S + O_2 \rightarrow SO_2$$

$$2SO_2 + O_2 \rightarrow 2SO_3$$

$$SO_3 + H_2O \rightarrow H_2SO_4$$

What mass of sulphuric acid be obtained from 3.20 g of sulphur ?

18. 5 g of an impure sample of sodium bicarbonate when heated strongly gave 600 mL of CO_2 measured at 27°C and 760 mm Hg pressure. Calculate the percentage purity of the sample.
19. A small piece of commercial zinc weighing 10 g is made to react with excess of dil. sulphuric acid. The total volume of hydrogen gas liberated was found to be 3.1 L at NTP. Determine the percentage purity of the zinc sample. Name the limiting reagent.
20. What weight of copper sulphate can be obtained by the action of 2.941 g of hot concentrated sulphuric acid on excess of copper. Identify the limiting reagent.

21. Calculate the percentage yield of oxygen if 0.0240 g of oxygen are produced from 0.303 g of KNO_3 on heating.

22. How many grams of diborane (B_2H_6) can be produced from 0.1 millimole of $LIAlH_4$ according to the reaction.

    $BF_3 + LiAlH_4 \rightarrow B_2H_6 + LiF + AlF_3$

    [Hint. Balance the equation first].

23. Balance the following equations and calculate the theoretical yield of the designated product from the quantity of the starting material given.

    (a) $BiS_2 + O_2 \rightarrow Bi_2O_3 + SO_2$

    16.5 g ?

    (b) $P_2I_4S_2 \rightarrow P_4S_7 + PI_3 + I_2$

    0.262 g ?

24. What was the percent yield of sodium sulphate which was made from 4.19 g of sodium carbonate, if 5.61 g was obtained ?

25. Calculate the atomic mass of calcium if 1.0395 g of $CaCl_2$ gave 2.6847 g of AgCl.

26. Name all the laws of chemical combination.

27. State the law of conservation of mass.

28. Why do mixtures not obey the law of constant composition?

29. Which of the following obeys the law of constant composition?

    (a) heterogeneous mixture

    (b) homogeneous mixture

    (c) a solution

    (d) an element

    (e) a compound

30. Which postulate of the Dalton's atomic theory needed modification

    (a) after the discovery of isotopes?

    (b) after the discovery of radioactivity?

31. How is an atom defined in the modern atomic theory?

32. Deduce the law of multiple proportions from the Dalton's theory.
33. State Avogadro's hypothesis.
34. Define molar volume of a substance.
35. How many molecules are there in one mole of a substance? By what name this number is called?
36. Define atomicity of a gas. What is the atomicity of
    (a) hydrogen (b) helium
    (c) chlorine (d) ozone?
37. Show that. Molecular mass = 2 × Vapour density.
38. Quicklime contains 71.47% of calcium. How much calcium is present in a sample of quicklime which contains 16 g of oxygen?
39. Which laws of chemical combination are illustrated by the following data:
    (a) An oxide which contains 63.2% of manganese?
    (b) For every 54.94 g of manganese in a compound, there is 64.00 g of oxygen?
40. 2.16 g of metallic copper was converted into nitrate by reacting with nitric acid. This nitrate on ignition gave 2.70 g of copper oxide. In another experiment, 1.15 g of copper oxide on reduction yielded 0.92 g of copper. Show that the results illustrate the law of definite proportions (or law of constant composition).
41. In an experiment, 0.32 g of sulphur on burning in air produces 224 mL of SO_2 at NTP. In another experiment, a metal sulphite reacts with a mineral acid to produce SO, gas which contains 50% sulphur. Show that the data illustrates the law of constant composition.
42. Phosphorous forms two common oxides. An early investigator was surprised to find that whereas one of these contained 56.35% of phosphorus by mass, the other one contained 56.35% of oxygen. Show that the data support the law of multiple proportions.
43. Copper gives two oxides. On heating 1 g of each in hydrogen, 0.888 g and 0.798 g of copper respectively were obtained. Show that the results are in agreement with the law of multiple proportion.

44. Two oxides of lead were separately reduced to metallic lead by heating in a current of hydrogen and the following data were obtained.

	Yellow oxide	Brown oxide
Mass of oxide	3.45g	1.195g
Loss of mass on reduction	0.24g	0.156g

Show that the data illustrates the law of multiple proportions.

45. From the following data, show that the chlorides of phosphorus illustrate the law of multiple proportions.

	1st compound	2nd compound
Mass % of phosphorous	22.54	14.88
Mass % of chlorine	77.46	85.12

46. From the data given below show that the law of constant composition and the law of multiple proportions are verified.

Sample	*Compound*	*Mass of carbon/g*	*Mass of oxygen/g*
1	Carbon dioxide	0.2002	0.5333
2	Carbon dioxide	0.1682	0.4480
3	Carbon monoxide	0.1543	0.2056
4	Carbon monoxide	0.5147	0.6856

47. Copper sulphide contains 66.5% copper, copper oxide contains 79.9% copper and sulphur trioxide contains 40% sulphur. Show that the data illustrates the law of reciprocal proportions.

48. 10 L of nitrogen gas and 10 L of hydrogen gas are introduced into an evacuated flask of 10 L capacity. It is then heated to 700 K and 3 atm pressure. What volume of ammonia (at NTP) is produced?

49. Define atomic mass unit.

50. In what respect do the atomic mass and relative atomic mass of a substance differ?

51. Define empirical formula. Illustrate by giving a suitable example.

52. Define the term mole.

53. How many particles are there in one mole of a substance?

54. How much volume one mole of a gaseous substance occupies at NTP?

55. Define molar mass. What is the unit of molar mass?

56. In a certain mass of a gas, the number of atoms and the number of molecules are equal. What conclusion can you draw from this observation?

57. Determine the molecular mass of $FeSO_4.(NH_4)_2SO_4.6H_2O$. For the atomic masses, please refer to table on atomic masses.

58. What is the number of water molecules contained in a drop of water weighing 0.06 g?

59. How may atoms are there in 16.0 g of $^{16}_{8}O$?

60. Calculate the mass of 1 atomic mass unit in grams.

61. Calculate the volume of 8 g of oxygen under NTP conditions.

62. Calculate the number of moles (n) in m g of a substance having molecular mass M u.

63. Calculate the number of moles of a gaseous substance present in V mL (at NTP) of it?

64. Calculate the mass of 5.5 billion molecules (nearly equal to the population on the Earth) of water.

    Atomic masses are: H = 1.0 u, and 0 = 16.0 u.

65. How many years would it take to spend Avogadro number of rupees at the rate of 10 lakh rupees per second?

66. Calculate the number of gold atoms in 0.3 g of 20 carat gold. The 24 carat gold is taken as 100% pure gold. Atomic mass of gold = 197 u.

67. Calculate the total number of electrons present in 1.6 g of methane.

68. How many ions each of Na^+ and Cl^- are present in 12.0 g of NaCl?

69. What do you mean by the term chemical formula of a compound?

70. Write the names of the various types of chemical formulae.

71. Define empirical and molecular formulae. How are the two related to each other?

72. Give one example each of a molecule in which the empirical formula and the molecular formula are, (a) the same (b) different.

73. Write down the empirical formulae of the compounds having the following molecular formulae.

(a) C_6H_{12} (b) Na_2CO_3
(c) B_2H_6 (d) N_2O_4
(e) H_2O_2 (f) H_2O
(g) C_6H_6

74. Calculate the percentage of copper in each of the following:

(a) Cuprite (Cu_2O) (b) Copper pyrites ($CuFeS_2$)
(c) Malachite ($CuCO_3Cu(OH)_2$)

How many tons of cuprite will give 500 tons of copper?

75. A compound was known to be either $Cucl_2$ or $CuBr_2$. A 5.0 g sample yields 2.36 g of copper upon reduction. Identify the compound.

[Hint. Calculate the percentage of copper in the sample and compare with that of $CuCl_2$ and $CuBr_2$]

76. A hydrocarbon contains 92.3% carbon and 7.66% of hydrogen. Its molecular mass is found to be 78 u. Determine its molecular formula.

77. A carbohydrate containing 40.0% C, 6.73%. H and 53.3% 0 has a molecular mass of 180.2 u. Determine its molecular formula.

78. A gaseous compound gave the following results on analysis. N = 9.48%, S = 20.9%, P = 38.0%. Determine the simplest formula of this compound.

79. An oxide of manganese was found to contain 69.64% manganese. What is the simplest formula of the oxide?

80. A compound is found to have the following composition:

Na = 14.31%, S = 9.97%, H = 6.22% and 0 = 69.50%.

If all the hydrogen in this compound is present as water of crystallisation, then determine the molecular formula of the compound? The molecular mass of the compound is 322 u. (Atomic masses are: Na = 23 u, S = 32 u, H = 1 u and 0 = 16 u).

81. Determine the empirical formula of a compound having the percentage composition

Fe = 20%, S = 11.5%, 0 = 23.1% and H_2O (water) = 20.35%.

Atomic masses (in amu) are: Fe = 56, S = 32, 0 = 16.

82. A complex metal organic compound was found to contain 60.3% carbon, 4.3% hydrogen, 33.4% tungsten, and the remainder oxygen. Its molecular mass was found to be 564 u. Determine its molecular formula?

83. Determine the formula of the compound from which the following data were obtained.

Mass of the compound/mg	H/mg	CH_2/mg	Al/mg
64.5	1.196	33.82	29.94

84. A mineral named antarcticite (because it is found from a small pond in Antarctica) gave the from data after analysis:

Ca = 17.5%, Cl = 32.7%, H_2O = 49.2%

What is the simplest formula of the compound?

85. Aspirin contains 60.0% C, 4.48% H, and the rest oxygen. Determine the empirical formula of aspirin.

86. What do you understand by the terms: precision and accuracy?

87. Comment upon the statement that "a precise measurement may not be accurate."

88. What is the international system of units? Name the seven dimensionally independent physical quantities recommended by IUPAC.

89. Define the physical quantity, amount of substance. What is its unit?

90. Derive the SI unit for each of the following physical quantities: force, pressure, electric charge, frequency, area, molality and molarity.

91. Rewrite the following in terms of SI prefixes: 5Å, 4.1×10^{-11} m, 6.023×10^{10} g, and 8.3×10^{7}rgs.

92. What is meant by a conversion factor? Obtain the conversion factor for converting kilometres into metres.

93. What do you mean by significant figures? State the rules followed in counting the number of significant figures in a measured quantity.

94. What is an exponential number? How is an exponential number expressed? How do we determine the number of significant figures in an exponential number?

95. What is matter, and how is it classified?

96. Define the terms: element, compound and mixture.

97. How do solutions differ from ordinary mixtures?

98. Classify the followings into element, compound and mixture— copper, gold, brass, common salt, air and sea water.

99. Show that air is a mixture not a chemical compound.

100. What are the characteristics of a compound?

101. How do mixtures differ from compounds?

102. Write a short note on (a) sublimation (b) filtration (c) distillation.

103. How would you separate the constituents of a mixture containing

    (a) iodine crystals and sand (b) sugar and water (c) acetone and methyl alcohol.

104. Write a short note on chromatography. What is meant by the mobile phase and the stationary phase?

105. What is the solvent extraction method? How can iodine be extracted from its very dilute solution in water?

106. Describe fractional distillation. Mention one commercial process based on fractional distillation.

107. How does ordinary distillation differ from fractional distillation?

108. When two substances A and B are mixed together in a pestle and mortar, a large amount of heat is liberated and a new substance C is formed. C has properties entirely different from those of A and B. What is the chemical nature of C?

109. Name the various laws of chemical combination. Explain with an example the law of constant proportions.

110. Describe a simple experiment to prove the law of conservation of mass.

111. Name the various laws of chemical combination. How do these laws support the Dalton's atomic theory?
112. Illustrate by giving suitable example, the law of definite proportions.
113. State and explain,
    (i) law of multiple proportions,
    (ii) law of reciprocal proportions,
    (iii) Gay-Lussac's law of gaseous volumes.
114. How is the Dalton's atom redefined in view of the discovery of isotopes, and the discharge tube experiments?
115. Derive the law of constant proportions from Dalton's atomic theory.
116. State the Avogadro's law. Derive it from the ideal gas equation.
117. Using Avogadro's law, derive the atomicity of hydrogen gas.
118. Using Avogadro's law derive a relationship between the molar mass of a substance and its vapour density.
119. With the help of Avogadro's law show that the molecular formula of hydrogen chloride is HCl.
120. Differentiate between molarity and molality.
121. What is the effect of temperature on, (a) molarity, (b) molality, (c) molefraction?
122. Explain the terms, (i) mass fraction, (ii) molefraction.
123. How does the term molarity differ from formality?
124. How does the molarity of a solution change with temperature?
125. Why does the molality of a solution remain unchanged with temperature?
126. Define molefraction. A 95 mass per cent aqueous solution of ethanol is further diluted with water. The molefraction of ethanol in the diluted solution is 0.25. What is the molefraction of water in this solution?
127. What is the molarity of acetic acid solution containing 6 g of acetic acid per litre of the solution?

128. What is the molefraction of methanol in solution containing 3.2 g of methanol in 18 g of water.

129. What is atomic mass unit? Calculate the absolute mass equivalent to 1 atomic mass unit.

130. Calculate the molefraction of ethyl alcohol and water in a solution in which 46 g of ethyl alcohol and 180 g of water are mixed together.

131. Calculate the molality of a solution of 36 g of glucose in 500 g of water.

132. If 4 g of NaOH are dissolved in 100 cm^3 of the solution, what shall be the difference in its normality and molarity?

133. The density of 3 M aqueous solution of sodium thiosulphate is 1.25 g/mL. Calculate, (i) molefraction of sodium thiosulphate, (ii) molalities of Na^+ and $S_2O_3^{2-}$ ions.

134. The percentage composition (by mass) of a solution is, 45% X, 15% Y, and 40% Z. Calculate the molefraction of each component of the solution. (Molecular mass: X = 18, Y = 60 and Z = 60).

135. 8.0575×10^{-2} kg of glauber's salt is dissolved in water to obtain 1 dm^3 of a solution of density 1077.2 kg m^{-3}. Calculate the molarity, molality, and molefraction of Na_2SO_4 in the solution.

   [Hints: Glauber's salt is $Na_2SO_4.10H_2O$ (molar mass = 322g/mol) and, 1 mol $Na_2SO_4.10H_2O$ = 1 mol of Na_2SO_4].

136. Define (a) atomic mass of an element, (b) molecular mass of a substance. What are the units of atomic and molecular masses?

137. Copper gives two oxides. On heating 1 g of each in hydrogen, 0.888 g and 0.798 g of copper respectively were obtained. Show that the results are in agreement with the law of multiple proportions.

138. Discuss mole concept. How many atoms are there in one mole of copper?

139. Explain the mole concept. Derive an expression which should give the number of molecules in terms of mass and molecular mass of the substance.

140. What is gram-molecular mass? How does it differ from molecular mass?

141. Define molar volume of a substance. What is the molar volume of a gas under NTP conditions?

142. Define empirical and molecular formulae. How are they related to each other?

143. What is the significance of an empirical formula of a compound? Write the empirical formula of diamond.

144. What does the molecular formula of a compound signify? Name a compound whose empirical formula is the same as its molecular formula.

145. Define a chemical equation. What is meant by a balanced chemical equation?

146. What are the essentials of a chemical equation?

147. What quantitative information does a balanced chemical equation convey?

148. Why should a chemical equation be balanced?

149. Describe briefly the different methods of balancing chemical equations.

150. Define the limiting reagent in a reaction. Illustrate by giving a suitable example.

151. Write some typical relationships on which the calculations based on chemical equations are based.

152. Name the various units used for describing concentration of a solute in any solution.

153. Name the units of concentration based on the molecular characteristic of the solute.

154. Define molarity and molality of a solution. Under what conditions, these two will have the same value?

155. Write the relationship for describing

(a) molarity

(b) normality

(c) molality

in terms of mass of solute, mass/volume of solvent / solution and the molar mass of the solute.

156. Which of the following are temperature independent?
   (a) mass percentage
   (b) volume percentage
   (c) molality
   (d) molarity
   (e) normality
   (f) mole fraction

157. Which of the following are particularly useful for
   (i) measuring the extent of chemical reactions of the solute,
   (ii) studying the reactions involving gases?
   (a) mass percentage
   (b) volume percentage
   (c) molefraction
   (d) molality

158. 3.2 g of sulphur combines with 3.2 g of oxygen to form a compound under one set of conditions. In another experiment, 0.8 g of sulphur combines with 1,2 g of oxygen to form another compound. State the law illustrated by these results.

159. Carbon combines with hydrogen to form three compounds A, B and C. The percentages of hydrogen in A, B and C are 25%, 14.3% and 7.7% respectively. Which law of chemical combination is illustrated by these data?

160. The oxides of a metal contain 27.6% and 30% oxygen respectively. If the formula of the first is Mg_3O_4, find that of the other.

161. Hydrogen sulphide contains 94.11% of sulphur, sulphur dioxide contains 50% of oxygen, and water contains 11.11% of hydrogen. Show that the data support the law of reciprocal proportions.

162. 50 ml of a gaseous mixture of hydrogen and hydrogen chloride was exposed to sodium amalgam. The volume decreased to 42.5 mL. If 100 mL of the same mixture is mixed with 50 mL of gaseous. ammonia and then exposed to water, what will be the final volume? All volumes being measured under similar conditions of temperature and pressure.

163. Calculate the number of molecules present in 1 cm^3 of hydrogen gas at NTP.

164. Calculate the number of atoms of zinc in a sample weighing 51.8 g.

165. How many mercury atoms are present in 36.0 cm3 of the element at 20°C.

166. Calculate the total number of electrons present in 1.6 g of methane.

167. What mass in kilogram of K_2O contains the same number of moles of K atoms as are present in 1.0 kg of KCl?

168. How many molecules of CO_2 are contained in one litre of air if the volume content of CO_2 in air in 0.03%?

169. A sample of copper chloride weighing 4.00 g was found to contain 1.89 g of copper and 2.11 g of chlorine. Determine its empirical formula.

170. Calculate the empirical formulae for the substances with the following analysis (by mass).

(a) C = 75.0%, H = 25.0%,

(b) H = 6.25%, N = 43.75%, O = 50.0%

(c) H = 1%, Cl = 35.3%, O = 63.7%

(d) Fe = 24.25%, C = 15.65%, N = 18.30%, S = 41.8%

(e) K = 26.55%, Cr = 35.35%, O = 38.10%

(f) C = 49.3%, H = 9.6%, N = 19.2%, O = 21.9%

171. What is the molecular formula for a compound whose molecular mass is 116 u, and whose empirical formula is CHO?

172. When 2.435 g of antimony is heated with excess sulphur, a chemical reaction occurs. The excess sulphur is driven off, leaving only the compound. If 3.397 g of compound are produced, what is the empirical formula of the compound?

173. An organic compound contains C = 31.58%, H = 5.26% and 0 = 63.16. Establish its empirical formula.

174. An organic compound was found to contain C = 40.65%; H = 8.55% and N = 23.7%. Its vapour density was found to be 29.5. Determine its empirical and molecular formulae.

175. A solution containing 2.68×10^{-2} moles of an ion A^{n+} require 1.61×10^{-3} moles of MnO_4^- for the oxidation of A^{n+} to AO_3^- in acidic medium. What is the value of n?

176. When 16.8 g of a white solid X was heated, 4.4 g acid gas A that turned lime-water milky was driven off together with 1.8 g of a gas B which condensed to a colourless liquid. The solid that remained, dissolved in water to give an alkaline solution, which with excess barium chloride solution gave a white precipitate Z. The precipitate effervesced with acid giving off carbon dioxide. Identify A, B and X. Write down the equation for the thermal decomposition of X.

177. When 20.0 g of a white solid X is heated, 4.4 g an acid gas A and 1.8 g of a neutral gas B are evolved, leaving behind a solid residue Y of weight 13.8 g. A turns limewater milky and B condenses into a liquid which changes anhydrous copper sulphate blue. The aqueous solution of Y is alkaline to litmus and gives 19.7 g of a white precipitate Z with barium chloride solution. Z gives carbon dioxide with an acid. Identify A, B, X, Y and Z.

178. Upon mixing 45.0 mL of 0.25 M lead nitrate solution with 25.0 mL of 0.10 M chromic sulphate, precipitation of lead sulphate takes place. How many moles of lead sulphate are formed? Also, calculate the molar concentration of the species left behind in the final solution. Assume that lead sulphate is completely insoluble.

179. Some sodium chloride is mixed in a sample of anhydrous sodium carbonate. On heating 12 g of this sample with dilute HCl, 2240 mL of CO_2 gas is obtained at N.T.P. Calculate the percentage of sodium chloride in the sample.